T0258536

Encyclopedia of Alternative and Renewable Energy: Novel Aspects of Solar Cells

Volume 29

Encyclopedia of Alternative and Renewable Energy: Novel Aspects of Solar Cells Volume 29

Edited by **Terence Maran and David McCartney**

New York

Published by Callisto Reference,
106 Park Avenue, Suite 200,
New York, NY 10016, USA
www.callistoreference.com

Encyclopedia of Alternative and Renewable Energy:
Novel Aspects of Solar Cells
Volume 29
Edited by Terence Maran and David McCartney

International Standard Book Number: 978-1-63239-203-9 (Hardback)

Printed in the United States of America.

Contents

Preface   VII

Chapter 1   **A New Guide to Thermally Optimized**
Doped Oxides Monolayer Spray-Grown Solar Cells:
The Amlouk-Boubaker Optothermal Expansivity ψ_{AB}   **1**
M. Benhaliliba, C.E. Benouis,
K. Boubaker, M. Amlouk and A. Amlouk

Chapter 2   **Effects of Optical Interference**
and Annealing on the Performance of
Polymer/Fullerene Bulk Heterojunction Solar Cells   **17**
Chunfu Zhang, Hailong You, Yue Hao,
Zhenhua Lin and Chunxiang Zhu

Chapter 3   **Flexible Photovoltaic Textiles for Smart Applications**   **43**
Mukesh Kumar Singh

Chapter 4   **Organic-Inorganic Hybrid Solar Cells:**
State of the Art, Challenges and Perspectives   **69**
Yunfei Zhou, Michael Eck and Michael Krüger

Chapter 5   **Dilute Nitride GaAsN and InGaAsN Layers**
Grown by Low-Temperature Liquid-Phase Epitaxy   **95**
Malina Milanova and Petko Vitanov

Chapter 6   **Relation Between Nanomorphology and**
Performance of Polymer-Based Solar Cells   **121**
Almantas Pivrikas

Chapter 7   **One-Step Physical Synthesis of Composite Thin Film**   **149**
Seishi Abe

Chapter 8   **Bioelectrochemical Fixation of Carbon**
Dioxide with Electric Energy Generated by Solar Cell   **167**
Doo Hyun Park, Bo Young Jeon and Il Lae Jung

Chapter 9   **Cuprous Oxide as an Active Material for Solar Cells**   191
Sanja Bugarinović, Mirjana Rajčić-Vujasinović,
Zoran Stević and Vesna Grekulović

Chapter 10   **Semiconductor Superlattice-Based
Intermediate-Band Solar Cells**   211
Michal Mruczkiewicz, Jarosław W. Kłos and Maciej Krawczyk

Permissions

List of Contributors

Preface

The main aim of this book is to educate learners and enhance their research focus by presenting diverse topics covering this vast field. This is an advanced book which compiles significant studies by distinguished experts in the area of analysis. This book addresses successive solutions to the challenges arising in the area of application, along with it; the book provides scope for future developments.

This book covers topics related to photovoltaic generations, its unique scientific ideas and technical solutions. Issues related to the capability of solar cells and the hydrogen productions in photoelectrochemical solar cells have also been discussed. Usage of advanced materials, like the oxide of copper as an active substance for solar cells, InP in solar cells with MIS arrangement, AlSb for use as an absorber layer, among others, have been discussed. Furthermore, other significant topics such as the analysis of both the status and prospects of organic photovoltaics such as polymer/fullerene solar cells, poly (p-phenylene-vinylene) derivatives, etc. have also been included.

It was a great honour to edit this book, though there were challenges, as it involved a lot of communication and networking between me and the editorial team. However, the end result was this all-inclusive book covering diverse themes in the field.

Finally, it is important to acknowledge the efforts of the contributors for their excellent chapters, through which a wide variety of issues have been addressed. I would also like to thank my colleagues for their valuable feedback during the making of this book.

Editor

A New Guide to Thermally Optimized Doped Oxides Monolayer Spray-Grown Solar Cells: The Amlouk-Boubaker Optothermal Expansivity ψ_{AB}

M. Benhaliliba[1], C.E. Benouis[1],
K. Boubaker[2], M. Amlouk[2] and A. Amlouk[2]
[1]*Physics Department, Sciences Faculty, Oran University of Sciences and Technology
Mohamed Boudiaf- USTOMB, POBOX 1505 Mnaouer- Oran,*
[2]*Unité de Physique des dispositifs à Semi-conducteurs UPDS,
Faculté des Sciences de Tunis, Campus Universitaire 2092 Tunis,*
[1]*Algeria*
[2]*Tunisia*

1. Introduction

PVC Photovoltaic solar cells are unanimously recognized to be one of the alternative renewable energy sources to supplement power generation using fossils. It is also recognized that semiconductors layered films technology, in reducing production costs, should rapidly expand high-scale commercialization.

Despite the excellent achievements made with the earliest used materials, it is also predicted that other materials may, in the next few decades, have advantages over these front-runners. The factors that should be considered in developing new PVC materials include:

- Band gaps matching the solar spectrum
- Low-cost deposition/incorporation methods
- Abundance of the elements
- Non toxicity and environmental concerns,

Silicon-based cells as well as the recently experimented polymer and dye solar cells could hardly fit all these conditions. Transparent conducting oxides as ZnO, SnO$_2$ as well as doped oxides could be good alternative candidates.

In this context, the optothermal expansivity is proposed as a new parameter and a guide to optimize the recently implemented oxide monolayer spray-grown solar cells.

2. Solar cells technologies and design recent challenges

In spite of better performance of traditional junction-based solar cells, during the past few decades, reports have appeared in literature that describe the construction of cells based metal-oxides (Bauer et al., 2001; Sayamar et al., 1994; He et al., 1999; Tennakone et al., 1999;

Bandara & Tennakone, 2001) and composite nanocrystalline materials (Palomares et al., 2003; Kay & Gratzel, 2002). Since that time, several other semiconductors have been tested with less success.

Recent challenges concerning newly designed solar cells are namely Band-gap concerns, cost, abundance and environmental concerns.

2.1 Band gaps matching the solar spectrum

The recently adopted layered structure of PVC raised the problem of solar spectrum matching (Fig.1) as well as lattice mismatch at early stages. In fact, the heterogeneous structure: Contact/window layer/buffer layer/Contact causes at least three differently structured surfaces to adhere under permanent constraints. It is known that the electronic band gap is the common and initial choice-relevant parameter in solar cells sensitive parts design. It is commonly defined as the energy range where no electron states exist. It is also defined as the energy difference between the top of the valence band and the bottom of the conduction band in semiconductors. It is generally evaluated by the amount of energy required to free an outer shell electron the manner it becomes a mobile charge carrier. Since the band gap of a given material determines what portion of the solar spectrum it absorbs, it is important to choose the appropriate compound matching the incident energy range. The choice of appropriated materials on the single basis of the electronic band gap is becoming controversial due the narrow efficient solar spectrum width, along with new thermal and mechanical requirements. It is rare to have a complete concordance between adjacent crystalline structures particularly in band gap sense.

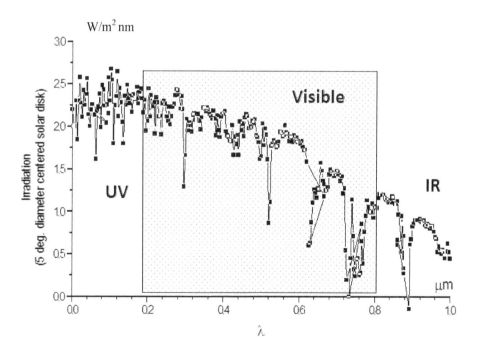

Fig. 1. Solar spectrum

A New Guide to Thermally Optimized Doped Oxides Monolayer Spray-Grown Solar Cells: The Amlouk-Boubaker
Optothermal Expansivity Ψ_{AB}

3

For example, in silicon-based solar cells, recombination occurring at contact surfaces at which there are dangling silicon bonds (Wu, 2005) is generally caused by material/phase discontinuities. This phenomenon limits cell efficiency and decreases conversion quality.

2.2 Low-cost deposition/incorporation methods
Deposition techniques and incorporation methods have been developed drastically and several deposition improved methods have been investigated for fabrication of solar cells at high deposition rates (0.9 to 2.0 nm/s), such as hot wire CVD, high frequency and microwave PECVD, , and expanding thermal plasma CVD. Parallel to these improvements, vacuum conditions and chemical processes cost increased the manner that serial fabrication becomes sometimes limited. Nowadays, it is expected that low processing temperature allow using a wide range of low-cost substrates such as glass sheet, polymer foil or metal. These features has made the second-generation low- cost metal-oxides thin-film solar cells promising candidates for solar applications.

2.3 Abundance of the elements
The first challenge for PV cells designer is undoubtedly the abundance of materials for buffer and window layers. The ratio of abundance i. e. of Tungsten-to-Indium is around 104, that of of Zinc-to-Tin is around 40. Although efficiency of Indium and Gallium as active doping agents has been demonstrated and exploited (Abe & Ishiyama, 2006; Lim et al., 2005), their abundance had decreased drastically (510 and 80 tons, respectively as reported by U.S. Geological Survey 2008) with the last decades' exploitation.

2.4 Non toxicity and environmental concerns
Among materials being used, cadmium junctions (Cd) and selenium (Se) are presumed to cause serious health and environmental problems. Risks vary considerably with concentration and exposure duration. Other candidate materials haven't gone though enough tests to show reassuring safety levels (Amlouk, 2010).

3. Materials optimisation

3.1 Primal selection protocols
Cost and toxicity concerns led to less and less use of Se and Cd-like materials. Additionally, increasing interest in conjoint heat-light conversion took some bad heat-conducting materials out from consideration. Selection protocols are becoming more concentrated on thermal, mechanical and opto-electric performance.
Since thermal conductivity, specific heat and thermal diffusivity has always been considered as material intrinsic properties, while absorbance and reflexivity depend on both material and excitation, there was a need of establishing advanced physical parameters bringing these proprieties together.

3.2 Opto-thermal analysis
The Amlouk-Boubaker optothermal expansivity is defined by:

$$\psi_{AB} = \frac{D}{\hat{\alpha}} \tag{1}$$

Where D is the thermal diffusivity and $\hat{\alpha}$ is the effective absorptivity, defined in the next section.

3.2.1 The effective absorptivity

The effective absorptivity $\hat{\alpha}$ is defined as the mean normalized absorbance weighted by $I(\tilde{\lambda})_{AM1.5}$, the solar standard irradiance, with $\tilde{\lambda}$: the normalised solar spectrum wavelength:

$$\begin{cases} \tilde{\lambda} = \dfrac{\lambda - \lambda_{min}}{\lambda_{max} - \lambda_{min}} \\ \lambda_{min} = 200.0 \text{ nm} \quad ; \lambda_{max} = 1800.0 \text{ nm}. \end{cases} \tag{2}$$

and :

$$\hat{\alpha} = \frac{\int\limits_0^1 I(\tilde{\lambda})_{AM1.5} \times \alpha(\tilde{\lambda}) d\tilde{\lambda}}{\int\limits_0^1 I(\tilde{\lambda})_{AM1.5} d\tilde{\lambda}} \tag{3}$$

where: $I(\tilde{\lambda})_{AM1.5}$ is the Reference Solar Spectral Irradiance.

The normalized absorbance spectrum $\alpha(\tilde{\lambda})$ is deduced from the Boubaker polynomials Expansion Scheme *BPES* (Oyedum et al., 2009; Zhang et al., 2009, 2010a, 2010b; Ghrib et al., 2007; Slama et al., 2008; Zhao et al., 2008; Awojoyogbe and Boubaker, 2009; Ghanouchi et al.,2008; Fridjine et al., 2009 ; Tabatabaei et al., 2009; Belhadj et al., 2009; Lazzez et al., 2009; Guezmir et al., 2009; Yıldırım et al., 2010; Dubey et al., 2010; Kumar, 2010; Agida and Kumar, 2010). According to this protocol, a set of m experimental measured values of the transmittance-reflectance vector: $\left(T_i(\tilde{\lambda}_i); R_i(\tilde{\lambda}_i)\right)\big|_{i=1..m}$

versus the normalized wavelength $\tilde{\lambda}_i\big|_{i=1..m}$ is established. Then the system (4) is set:

$$\begin{cases} R(\tilde{\lambda}) = \left[\dfrac{1}{2N_0} \sum_{n=1}^{N_0} \xi_n \times B_{4n}(\tilde{\lambda} \times \beta_n) \right] \\ T(\tilde{\lambda}) = \left[\dfrac{1}{2N_0} \sum_{n=1}^{N_0} \xi_n' \times B_{4n}(\tilde{\lambda} \times \beta_n) \right] \end{cases} \tag{4}$$

where β_n are the 4n-Boubaker polynomials B_{4n} minimal positive roots (N_0 is a given integer and ξ_n and ξ_n' are coefficients determined through Boubaker Polynomials Expansion Scheme BPES.

Finally, the normalized absorbance spectrum $\alpha(\tilde{\lambda})$ is calculated using the relation (5) :

$$\alpha(\tilde{\lambda}) = \frac{1}{d\sqrt{2}} \sqrt{\left(\ln \frac{1 - R(\tilde{\lambda})}{T(\tilde{\lambda})} \right)^2 + \left(\ln \frac{(1 - R(\tilde{\lambda}))^2}{T(\tilde{\lambda})} \right)^2} \tag{5}$$

where d is the layer thickness.

The effective absorptivity $\hat{\alpha}$ is calculated using (Eq. 3) and (Eq. 5).

3.2.2 The Optothermal expansivity ψ_{AB}

The Amlouk-Boubaker optothermal expansivity unit is m³s⁻¹. This parameter, as
calculated in Eq. (1) can be considered either as the total volume that contains a fixed
amount of heat per unit time, or a 3D expansion velocity of the transmitted heat inside the
material.

3.2.3 The optimizing-scale 3-D Abacus

According to precedent analyses, along with the definitions presented in § 3.2, it was
obvious that any judicious material choice must take into account simultaneously and
conjointly the three defined parameters: the band gap E_g, Vickers Microhardness $H\upsilon$ and
The Optothermal Expansivity ψ_{AB}. The new 3D abacus (Fig. 2) gathers all these parameters
and results in a global scaling tool as a guide to material performance evaluation.

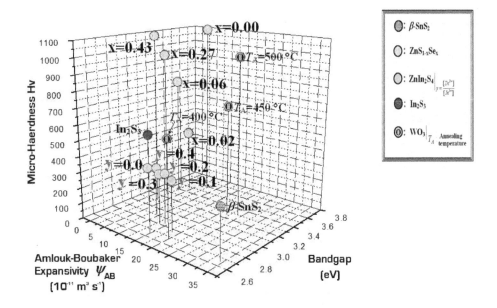

Fig. 2. The 3D abacus

For particular applications, on had to ignore one of the three physical parameters gathered
in the abacus. The following 2D projections have been exploited:
The projection in $H\upsilon$ - E_g plane, which is interesting in the case of a thermally neutral
material.
It is the case, i.e. of the ZnS$_{1-x}$Se$_x$ compounds, it is obvious that the consideration of Band
gap-Haredness features is mor important than thermal proprieties. The E_g - $H\upsilon$ projection
(Fig. 3) gives relevant information: the selenization process causes drastical loss of hardness
in initially hard binary Zn-S material.

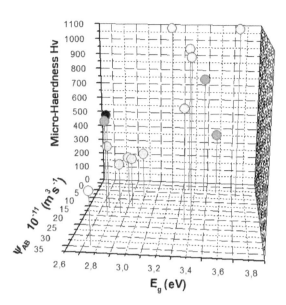

Fig. 3. The 3D abacus (E_g - H_υ projection)

This projection in ψ_{AB} - E_g plane is suitable for thick layers whose mechanical properties don't contribute significantly to the whole disposal hardness.

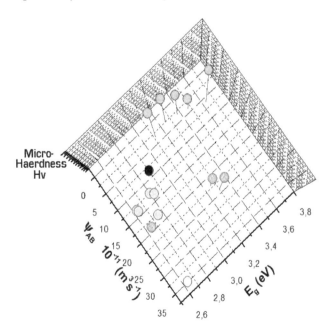

Fig. 4. The 3D abacus (ψ_{AB} - E_g projection)

The projection in ψ_{AB}-Hυ plane is useful for distinguishing resistant and good heat conductor materials, which is the case of the ZnIn$_2$S$_4$ materials.

In fact the effect of the Zinc-to-Indium ratio on the values of the Amlouk-Boubaker optothermal expansivity (Fig. 5) is easily observable in this projection (it is equivalent to an expansion of the values of the parameter ψ_{AB} into a wide range: [10-20] 10^{-11} m^{3}s^{-1}).

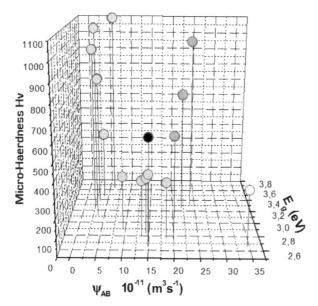

Fig. 5. The 3D abacus (ψ_{AB} - Hυ projection)

3.3 Investigation of the selected materials

According to the information given by the 3D abacus (Figures 3-5), some materials have been selected. ZnO and ZnO-doped layered materials, SnO$_2$ and SnO$_2$:F/SnO$_2$:F-SnS$_2$ compounds were among the most interesting ones.

3.3.1 ZnO and ZnO-doped layers

Zinc oxide (ZnO) is known as one of the most multifunctional semiconductor material used in different areas for the fabrication of optoelectronic devices operating in the blue and ultra-violet (UV) region, owing to its direct wide band gap (3.37 eV) at room temperature and large exciton binding energy (60 meV) (Coleman & Jagadish, 2006). On the other hand, it is one of the most potential materials for being used as a TCO because of its high electrical conductivity and high transmission in the visible region (Fortunato et al., 2009).

Zinc oxide can be doped with various metals such as aluminium (Benouis et al., 2007) indium (Benouis et al., 2010), and gallium (Fortunato et al., 2008). The conditions of deposition and the choice of the substrate are important for the growth of the films (Benhaliliba et al., 2010). The substrate choosen must present a difference in matching lattice less than 3% to have good growth of the crystal on the substrate (Teng et al., 2007; Romeo et al., 1998). ZnO (both doped and undoped) is currently used in the copper indium gallium diselenide (CIGS, or Cu (In, Ga)Se2) thin-film solar cell (Wellings et al., 2008; Haung et al., 2002). ZnO is also promising for the application in the electronic and sensing devices, either as field effect transistors (FET), light sensor, gas and solution sensor, or biosensor.

In addition to its interesting material properties motivating research of ZnO as semiconductor, numerous applications of ZnO are well established. The world usage of ZnO in 2004 was beyond a million tons in the fields like pharmaceutical industry (antiseptic healing creams, etc.), agriculture (fertilizers, source of micronutrient zinc for plants and animals), lubricant, photocopying process and anticorrosive coating of metals.

In electronic engineering, Schottky diode are the most known ZnO-based unipolar devices. The properties of rectifying metal contacts on ZnO were studied for the first time in the late 60ties (Mead, 1965; Swank, 1966; Neville & Mead, 1970) while the first Schottky contacts on ZnO thin films were realized in the 80ties (Rabadanov et al., 1981; Fabricius et al., 1986).

The undoped and doped ZnO films grow with a hexagonal würtzite type structure and the calculated lattice parameters (a and c) are given in Table 1 (Benhaliliba et al. 2010).

Nature	Grain Size (Å)	Int. (%)	d (Å)	2θ (°)	Angle Shift (°)	TC	a (Å)	c (Å)	$(c-c_0)/c_0(\times10^{-5})$
Undoped									
(100)	217	6.3	2.81	31.78	0.009	0.50			-61.4
(002)	358	25.7	2.60	34.44	-0.019	2.33	3.24	5.20	
(101)	254	19.4	2.47	36.24	-0.008	1.67			
IZO									
(100)	239	100	2.81	31.80	-0.050	2.24			-3.84
(002)	211	53.5	2.60	34.42	-0.019	1.19	3.24	5.20	
(101)	195	85.5	2.47	36.28	-0,028	1.95			
AZO									
(100)	206	70.7	2.81	31.80	-0.011	1.52			-115.23
(002)	225	70.5	2.60	34.46	-0.039	1.48	3.24	5.20	
(101)	195	100	2.47	36.28	-0.028	2.13			

Table 1.

Many significant differences were observed for the undoped, Al- and In-doped ZnO thin films. The films with low thickness (150 nm) have a random orientation with several peaks as reported by Wellings et al. (2008), Ramirez et al. (2007) and Abdullah et al. (2009). The same kind of growth was obtained by Tae et al. (1996) for 150 nm thick films. Whereas on FTO, the predominant ZnO film grew to a thickness of 200-300 nm as stated by Schewenzer et al. (2006). Figures (6-8) give some information about some information about ZnO and ZnO-doped layers.

A New Guide to Thermally Optimized Doped Oxides Monolayer Spray-Grown Solar Cells: The Amlouk-Boubaker
Optothermal Expansivity Ψ_{AB}

9

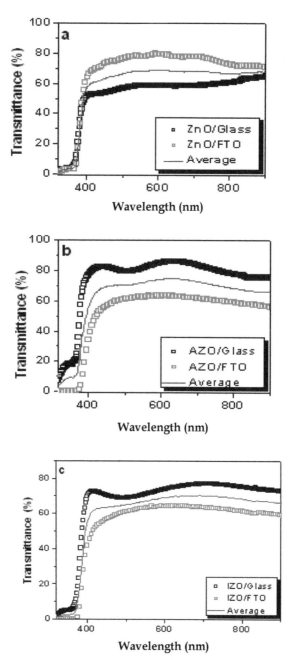

Fig. 6. Transmittance spectra, ZnO/Glass and ZnO/FTO (a), AZO/Glass and AZO/FTO
(b), IZO/Glass and IZO/FTO (c).

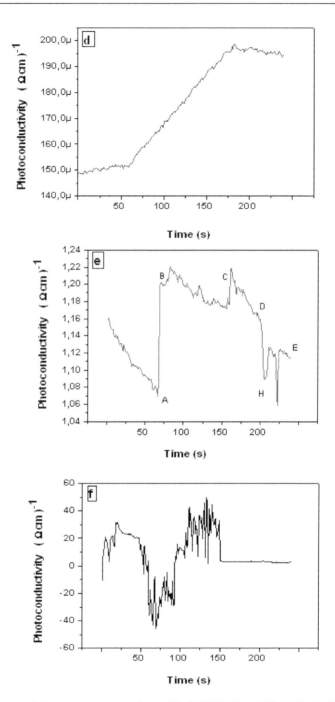

Fig. 7. Photoconductivity spectra versus time of ZnO/FTO (d), AZO/FTO (e), IZO/FTO (f).

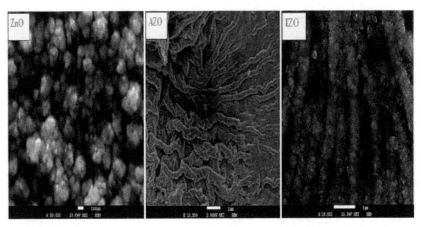

Fig. 8. SEM micrographs for (a) ZnO, (b) AZO and (c) IZO films, (bottom) white horizontal dashes indicate the scale (100 nm (ZnO), 1µm (AZO and IZO).

3.3.2 SnO₂:F-SnS₂ gradually grown layers

Tin oxide (SnO_2) is an n-type VI.II oxide semiconductor with a wide band gap ($Eg = 3.6$ eV). Because of its good opto-electrical properties, and its ability to induce a high degree of charge compensation, it is widely used as a functional material for the optoelectronic devices, gas sensor, ion sensitive field effect transistors, and transparent coatings for organic light emitting diodes (Onyia & Okeke, 1989; Wang et al., 2006; Lee & Park, 2006; Yamada et al., Kane & Schweizer,1976).

In the last decades, pure and doped tin oxide compounds, prepared by several techniques (Manorama et al., 1999; Bruno et al., 1994; Brinzari et al., 2001; Wang et al., 2002) have been used for the preparation of high performance gas sensing and light emitting devices layers (Barsan, 1994; Goepel & Schierbaum, 1995; Ramgir et al. ,2005).

SnO_2 thin films are generally prepared using methanol CH_4O: 1.0 L, demineralised water and anhydrous tin tetrachloride $SnCl_4$. Formation of pure SnO_2 is resulting from the endothermic reaction:

$$SnCl_4 + 2H_2O \xrightarrow{\;methanol,440°C\;} SnO_2 + 4\overset{\frown}{HCl}$$

Approximately 0.9 µm-thick SnO_2 thin films are generally deposited on glass, under an approximated substrate temperature T_s=440°C.

XRD patterns of the as-grown SnO_2 films are shown in Fig. 9. Diagram analysis shows that the layers present a first set of (110)-(101)-(200) X-ray diffraction peaks followed by more important pair (211)-(301). According to JCDPS 88-0287 (2000) standards, these patterns refer to tetragonal crystalline structure.

It was reported by Yakuphanoglu (2009) and Khandelwal et al. (2009)that SnO_2 films structure depends wholly on elaboration technique, substrate material and thermal treatment conditions. This feature was also discussed by Purushothaman et al. (2009) and Kim et al. (2008) who presented temperature-dependent structure alteration of the SnO_2 layers.

Atomic force microscopy (AFM) 3D images of the SnO_2 are presented in Fig. 10.

The layers present a pyramidal-clusters rough structure, which is characteristic to many Sn-like metal oxides. This observation confirms the XRD·results.

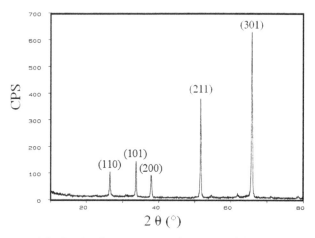

Fig. 9. XRD Diagram of SnO_2 thin layers prepared at T_s 440 °C.

Fig. 10. SnO_2 layers 3D and 2D surface topography 2D (top) and 3D (bottom).

SnO_2:F-SnS_2 gradually grown layers have as intermediate precursors SnO2:F layers obtained by spray pyrolysis on glass substrates according to the coupled reactions :

$$(SnCl_4, 5H_2O)_n + 2nH_2O \xrightarrow[\text{Carrier gas}]{\text{Spray}} n\,SnCl_4 + 7n\,H_2O\uparrow$$

A New Guide to Thermally Optimized Doped Oxides Monolayer Spray-Grown Solar Cells: The Amlouk-Boubaker
Optothermal Expansivity Ψ_{AB}

13

and

$$n\, SnCl_4 + n(NH_4^+, F^-) + 2n\, H_2O \xrightarrow[methanol]{440°C} n\, SnO_2{:}F + 4n\, \overrightarrow{HCl} + n\, \overrightarrow{NH_3} + \frac{n}{2}\, \overrightarrow{H_2}$$

In the second reaction, ammonium florid acts on the deposited (and heated) tin tetrachloride by incorporation process due to ionic close electro-negativity and dimension (F- and O²- radii ratio is around 0.96). The obtained layers are n-type (Fig 11-a)

Hence, the first step of the protocol is indeed elaboration of the precursor SnO_2: F layer. In the second step, this layer is subjected to local annealing in a highly sulfured atmosphere (Fig 11-b). Under specific experimental conditions (Temperature, pressure, exposure time) SnS_2 compound appears selectively at the top of the precursor SnO_2: F layer. This obtained mini-layer is n-type (fig 11-b).

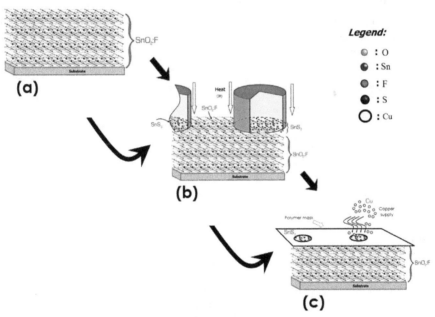

Fig. 11. TCO monolayer-grown: cell elaboration protocol

Finally, a neutral masking sheet is applied to the free surface in order to deposit copper (Cu) by evaporation, controlled dipping or even direct mechanical spotting. Due to the metallic diffusive properties, a multiphase CuSnS (Cu_2SnS_3, Cu_3SnS_4, Cu…) conducting compound appears at the free surfaces (Fig 11-c). This compound has been verified to have better mechanical performance than CuInS.

3.3.3 A sketch of the thermally optimized new monolayer grown cell
The first prototype of the proposed TCO monolayer-grown Solar cell is presented in Figure 12. The procedure can be applied to other oxides, namely Sb_xO_y, Sb_xS_y/MSbO (M=Cu, Ag,..) hetero-junction.

It has been experimented that n-type can be locally and partially transformed into p-WS_2, which results in a WO_3/WS_2 heterojunction, using the same sulfuration procedure detailed above.

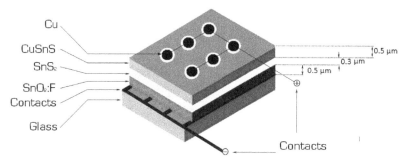

Fig. 12. TCO monolayer-grown Solar cell

The case of ZnO has been experimented but raised some problems, in fact it has been recorded that sulfuration process is never complete, and that an unexpected mixture $(ZnO)_x(ZnS)_y$ takes place.

4. Conclusion

In this chapter, a new physical parameter has been proposed as a guide for optimizing the recently implemented oxide monolayer spray-grown solar cells. This parameter led to the establishment of a 3D (bangap E_g -Vickers Microhardness $H\upsilon$ - Optothermal Expansivity ψ_{AB}) abacus. Thanks to optimizing features, some interesting materials have been selected for an original purpose: The TCO monolayer-grown Solar cell. The first prototype of the proposed TCO monolayer-grown Solar cell has been presented and commented. The perspective of using other oxides, namely Sb_xO_y, $Sb_xS_y/MSbO$ (M=Cu, Ag,..) has been discussed.

5. References

Abdullah, H.N.P.Ariyanto, S.Shaari, B.Yuliarto and S.Junaidi, Am. J. Eng. and Appl. Sc. 2 (2009) 236-240.

Abe, Y. & Ishiyama N., (2006). Titanium-doped indium oxide films prepared by DC magnetron sputtering using ceramic target. J. Mater. Sci. 41, pp.7580-7584

Agida, M., Kumar, A. S., 2010. A Boubaker Polynomials Expansion Scheme Solution to Random Love's Equation in the Case of a Rational Kernel , J. of Theoretical Physics 7,319.

Amlouk, A.; Boubaker K.& Amlouk M., (2010). J. Alloys Compds, 490,pp. 602–604.

Awojoyogbe, O.B., Boubaker, K., 2008. A solution to Bloch NMR flow equations for the analysis of homodynamic functions of blood flow system using m- Boubaker polynomials. Curr. Appl. Phys. 9, 278–283.

Bandara, J. & Tennakone, K. J. (2001). Colloid Interface Sci. 236, pp. 375-382.

Barsan, N. Sens. Actuators, B, Chem. 17 (1994) 241.

Bauer, C.; Boschloo, G., Mukhtar, E. & Hagfeldt, A. (2001). J. Phys. Chem. B 105,pp. 5585-5591.

Belhadj, A., Onyango, O., Rozibaeva, N., 2009. Boubaker polynomials expansion scheme-related heat transfer investigation inside keyhole model. J. Thermo- phys. Heat Transfer 23, 639–640.

Benhaliliba, C. E. Benouis, M. S. Aida, F. Yakuphanoglu, A. Sanchez Juarez, J Sol-Gel Sci
Technol (2010) 55:335–342 DOI 10.1007/s10971-010-2258-x.
Benhaliliba, C.E. Benouis, M.S. Aida, A. Sanchez Juarez, F. Yakuphanoglu, A. Tiburcio
Silver, J. Alloys Compd. 506 (2010) 548-553.
Benouis, C.E. ; Benhaliliba, M. , Sanchez Juarez, A., Aida, M.S., F.Yakuphanoglu, F., (2010).
Journal of Alloys and Compounds 490, pp. 62–67.
Benouis C.E., M. Benhaliliba, F. Yakuphanoglu, A. Tiburcio Silver, M.S. Aida, A. Sanchez
Juarez, Synthetic Metals (2011). D.O.I. 10.1016/ J. Synthmet.2011.04.017.
Benouis, C.E. ; Sanchez-Juarez, A., Aida, M.S., (2007). Phys. Chem. News, 35, pp. 72-79.
Brinzari V., G. Korotcenkov, V. Golovanov, Thin Solid Films 391 (2001) 167.
Bruno, L.C. Pijolat, R. Lalauze, Sens. Actuators, B, Chem. 18–19 (1994) 195.
Coleman, V. A. & Jagadish C., (2006). Basic Properties and Applications of ZnO, and:
chapter1 Zinc Oxide Bulk, Thin Films and Nanostructures C. Jagadish and S.
Pearton Elsevier limited.
Dubey, B., Zhao, T.G., Jonsson, M., Rahmanov, H. 2010. A solution to the accelerated-
predator-satiety Lotka–Volterra predator–prey problem using Boubaker
polynomial expansion scheme. J. Theor. Biology 264, 154-160.
Fabricius H, Skettrup T, Bisgaard P. Appl Opt 1986;25:2764–7.
Fortunato, E.; Gonçalves, A., Pimentel, A., Barquinha, P., Gonçalves, G., Pereira, L., Ferreira,
I. & Martins, R., (2009). Appl. Phys. A Mat. Sci. Proc. 96, pp.197-205.
Fortunato, E.; Raniero, L., Silva, L,. Gonçalves, A., Pimentel, A., Barquinha, P., Aguas, H.,
Pereira, L., Gonçalves, G., Ferreira, I., Elangovan, E., Martins, R., (2008). Sol. En.
Mat. And Sol. Cells 92, pp.1605-1610.
Fridjine, S., Amlouk, M., 2009. A new parameter: an ABACUS for optimizing functional materials
using the Boubaker polynomials expansion scheme. Mod. Phys. Lett. B 23, 2179–2182.
Ghanouchi, J., Labiadh, H., Boubaker, K., 2008. An Attempt to solve the heat transfer
equation in a model of pyrolysis spray using 4q-order Boubaker polynomials. Int. J.
Heat Technol. 26, 49–53.
Ghrib, T., Boubaker, K., Bouhafs, M., 2008. Investigation of thermal diffusivity–
microhardness correlation extended to surface-nitrured steel using Boubaker Ginot,
V. & Hervé, J. C. ,1994, Estimating the parameters of dissolved oxygen dynamics in
shallow ponds, Ecol. Model. 73, 169-187.
Goepel, W. Schierbaum, K.D. Sens. Actuators, B, Chem. 26 (1995) 1.
Guezmir, N., Ben Nasrallah, T., Boubaker, K., Amlouk, M., Belgacem, S., 2009. Optical
modeling of compound CuInS2 using relative dielectric function approach and
Boubaker polynomials expansion scheme BPES. J. Alloys Compd. 481, 543–548.
Haung F.J, Rudmann, D., Bilger, G., Zogg, H., Tiwari, A.N., (2002) Thin Solid Films 403-404,
pp. 293-296.
He, J.; Lindstrom, H., Hagfeldt, A. & Lindquist, S.E. (1999). J. Phys. Chem. B 103, pp. 8940-8951.
Kane, J. H.P. Schweizer, J. Electrochem. Soc. 123 (1976) 270.
Kay, A. & Gratzel, M. (2002). Chem. Mater 14, pp. 2930-2938.
Khandelwal, R. A. Pratap Singh, A. Kapoor, S. Grigorescu, P. Miglietta, N. Evgenieva
Stankova, A. Perrone, Optics & Laser Tech. 41 (2009) 89
Kim, H. H. Park, H. J. Chang, H. Jeon, H. H. Park, Thin Solid Films, 517(2008) 1072
Kumar, A. S., 2010. An analytical solution to applied mathematics-related Love's equation using the
Boubaker Polynomials Expansion Scheme, Journal of the Franklin Institute 347, 1755.
Lee, S.Y. B.O. Park, Thin Solid Films 510 (2006) 154.
Lim, J. H.; Yang E.-J., Hwang D.K., Yang J.H. (2005). Highly transparent and low resistance
gallium-doped indium oxide contact to p-type GaN. Appl. Phys. Lett. 87,pp. 1-3

Manorama, S.V. C.V.G. Reddy, V.J. Rao, Nanostruct. Mater. 11 (1999) 643.

Mead CA. Phys Lett (1965). Pp.18-218.

Nasr, C.; Kamat, P.V.& Hotchandani, S. J. (1998). Phys. Chem. B 102, pp.10047-10052.

Neville RC, Mead CA. J Appl Phys 1970;41:3795.

Onyia, A.I. C.E. Okeke, J. Phys. D: Appl. Phys. 22 (1989) 1515.

Oyodum, O.D., Awojoyogbe, O.B., Dada, M., Magnuson, J., 2009. On the earliest definition of the Boubaker polynomials. Eur. Phys. J. Appl. Phys. 46, 21201–21203.

Palomares, E.; Cliford, J.N., Haque, S.A., Lutz, T.& Durrant, J.R. (2003). J. Am. Chem. Soc 125, pp. 475-481.

Purushothaman, K.K. M. Dhanashankar, G. Muralidharan, Current App. Physics 9 (2009) 67

Rabadanov RA, Guseikhanov MK, Aliev IS, Semiletov SA. Fizika (Zagreb)1981;6:72.

Ramgir, N.S. Mulla, I.S. Vijayamohanan, K.P. J. Phys. Chem., B 109 (2005) 12297.

Ramirez, D.D.Silva, H.Gomez, G.Riveros, R.E.Marotti, E.D.Dalchiele, Solar Energy Materiels and Solar Cells 31 (2007) 1458-1451.

Redmond, G.; Fitzmaurice, D. & Gratzel, M. (1994). Chem. Mater. 6, pp. 686-689.

Romeo, A.; Tiwari, A.N., & Zogg, H., 2nd World Conference and Exhibition on Photovoltaic Solar Energy Conversion 6-10 July 1998 Hofburg Kongresszentru, Vienna Austria.

Sayama, K.; Sugihara, H. & Arakawa H.(1998). Chem. Mater. 10, 3825-3830.

Schewenzer, B.J.R.Gommm and D.E.Morse, Langmuir 22 (2006) 9829-9831.

Slama, S., Bessrour, J., Bouhafs, M., BenMahmoud, K. B., 2009a .Numerical distribution of temperature as a guide to investigation of melting point maximal front spatial evolution during resistance spot welding using Boubaker polynomials. Numer. Heat Transfer Part A 55,401–408.

Slama, S., Boubaker, K., Bessrour, J., Bouhafs, M., 2009b. Study of temperature 3D profile during weld heating phase using Boubaker polynomials expansion. Thermochim. Acta 482, 8–11.

Slama, S., Bouhafs, M., Ben Mahmoud, K.B., Boubaker, A., 2008. Polynomials solution to heat equation for monitoring A3 point evolution during resistance spot welding. Int. J. Heat Technol. 26, 141–146.

Swank RK. Phys Rev 1966;153:844.

Tabatabaei, S., Zhao, T., Awojoyogbe, O., Moses, F., 2009. Cut-off cooling velocity profiling inside a keyhole model using the Boubaker polynomials expansion scheme. Heat Mass Transfer 45, 1247–1251.

TaeYoung Ma, Sang Hyun Kim, Hyun Yul Moon, Gi Cheol Park, Young Jin Kim, Ki Wan Kim, J.Appl.Phys. 35(1996) 6208-6211.

Teng, X.; Fan, H., Pan, S., Ye, C., Li, G., (2007). Materials letters 61, pp. 201-204.

Tennakone, K.; Kumara, G. R. R. A., Kottegoda, I. R. M. & Perera, V.P.S. (1999). J. Chem. Soc. Chem. Commun. 99, pp. 15-21.

Wang, H.C.Y. Li, M.J. Yang, Sens. Actuators B 119 (2006) 380.

Wang, Y. C. Ma, X. Sun, H. Li, Nanotechnology 13 (2002) 565.

Wellings, J.S.; Chaure, N.B., Heavens, S.N., Dharmadasa, I.M., (2008). Thin Solid Films, 516, pp. 3893-3898.

Wellings, J.S.N.B.Chaure, S.N.Heavens, I.M.Dharmadasa, Thin Solid Films, 516 (2008) 3893-3898.

Wu, L. Z.; Tian W.& Jiang X. T. (2005). Silicon-based solar cell system with a hybrid PV module. Solar Energy Materials and Solar Cells. 87, pp.637-645

Yakuphanoglu, F. Journal of Alloys and Compounds, 470 (2009) 55

Yamada Y., K. Yamashita, Y. Masuoka, Y. Seno, Sens. Actuators B 77 (2001) 12.

Zhao, T.G., Wang, Y.X., Ben Mahmoud, K.B., 2008. Limit and uniqueness of the Boubaker–Zhao polynomials imaginary root sequence. Int. J. Math. Comput. 1, 13–16.

Effects of Optical Interference and Annealing on the Performance of Polymer/Fullerene Bulk Heterojunction Solar Cells

Chunfu Zhang[1], Hailong You[1], Yue Hao[1],
Zhenhua Lin[2] and Chunxiang Zhu[2]
[1]School Of Microelectronics, Xidian University,
[2]ECE, National University of Singapore,
[1]China
[2]Singapore

1. Introduction

Polymer solar cells are of tremendous interests due to their attractive properties such as flexibility, ease of fabrication, low materials and energy budget. However, organic materials have short exciton diffusion length and poor charge mobility, which can greatly decrease the performance of polymer solar cells. These challenges can be effectively overcome through the use of the bulk heterojunction (HJ) structure because it can guarantee the effective exciton dissociation and carrier transport simultaneously if a proper bicontinuous interpenetrating network is formed in the active layer. Based on this structure, the performance of polymer solar cells has been improved steadily in the past decade.

The performance of a polymer solar cell is mainly determined by the short-circuit current density (J_{SC}), the open circuit voltage (V_{OC}), and the fill factor (FF), given that $\eta = J_{SC}V_{OC}FF/P_{in}$ (where η is power conversion efficiency, PCE, and P_{in} is the incident optical power density). V_{OC} has a direct relationship with the offset energies between the highest occupied molecular orbital of Donor (D) material and the lowest unoccupied molecular orbital of Acceptor (A) material (Cheyns et al., 2008). Since the D and A materials are intimately mixed together in the bulk HJ structure and their interfaces distribute everywhere in the active layer, it is difficult to increase V_{OC} by changing D/A interface property for a given material system (such as poly(3-hexylthiophene-2,5-diyl):[6,6]-phenyl C_{61} butyric acid methyl ester, P3HT:PCBM). Thus the usually used optimization method is to improve J_{SC} and FF.

J_{SC} greatly depends on the optical interference effect in polymer solar cells. Because of the very high optical absorption ability of organic materials, the active layer is very thin and typically from several ten to several hundred nanometers. This thickness is so thin compared to the incident light wavelength that the optical interference effect has to be carefully considered. Depending on the thicknesses and optical constants of the materials, the optical interference causes distinct distributions of the electric field and energy absorption density. Due to this effect, J_{SC} shows an obvious oscillatory behavior with the variation of active layer thickness. In order to gain a high PCE, the active layer thickness needs to be well optimized according to the optical interference.

Besides the serious optical interference effect, J_{SC} also suffers from the non-ideal free carrier generation, low mobility and short carrier lifetime. In order to reduce the exciton loss and guarantee the efficient carrier transport, the optimal interpenetrating network, or to say, the optimal morphology is desired in the bulk HJ structure. In order to achieve an optimal morphology, a thermal treatment is usually utilized in the device fabrication, especially for the widely used P3HT:PCBM solar devices. It is found that the sequence of the thermal treatment is critical for the device performance (Zhang et al., 2011). The polymer solar cells with the cathode confinement in the thermal treatment (post-annealed) show better performance than the solar cells without the cathode confinement in the thermal treatment (pre-annealed). The functions of the cathode confinement are investigated in this chapter by using X-ray photoelectron spectroscopy (XPS), atomic force microscopy (AFM), optical absorption analysis, and X-ray diffraction (XRD) analysis. It is found that the cathode confinement in the thermal treatment strengthens the contact between the active layer and the cathode by forming Al–O–C bonds and P3HT-Al complexes. The improved contact effectively improves the device charge collection ability. More importantly, it is found that the cathode confinement in the thermal treatment greatly improves the active layer morphology. The capped cathode effectively prevents the overgrowth of the PCBM molecules and, at the same time, increases the crystallization of P3HT during the thermal treatment. Thus, a better bicontinuous interpenetrating network is formed, which greatly reduces the exciton loss and improves the charge transport capability. Meanwhile, the enhanced crystallites of P3HT improve the absorption property of the active layer. All these aforementioned effects together can lead to the great performance improvement of polymer solar cells. Besides the thermal treatment sequence, temperature is another very important parameter in the annealing process. Various annealing temperatures have also been tested to find the optimized annealing condition in this chapter.

The contents of this chapter are arranged as the following: Section 2 introduces the effects of the optical interference on J_{SC} in polymer solar cells by considering the non-ideal free carrier generation, low mobility and short carrier lifetime at the same time; Section 3 investigates the influence of the sequence of the thermal treatment on the device performance with emphasis on the cathode confinement in the thermal treatment; based on the optical interference study and the proper thermal treatment sequence, the overall device optimization is presented in Section 4. At last, a short conclusion is given in Section 5.

2. Effects of optical interference on J_{SC}

J_{SC} is directly related to the absorption ability of organic materials. It is believed that increasing the light harvesting ability of the active layer is an effective method to increase J_{SC}. In order to increase J_{SC}, some optical models (Pettersson, 1999; Peumans et al., 2003) have been built to optimize the active layer thickness. However, only optimizing the thickness for better light absorption is difficult to improve J_{SC}. This is because that PCE depends not only on the light absorption, but also on exciton dissociation and charge collection. In polymer solar cells, a blend layer consisting of conjugated polymer as the electron donor and fullerene as the electron acceptor is always used as the active layer. For a well blended layer, the length scale of D and A phases is smaller than the exciton diffusion length (typically less than 10 nm), so that most of the generated excitons can diffuse to the D/A interface before they decay. Even if all the excitons can reach the D/A interface, not all of them can be dissociated into free carriers. The exciton-to-free-carrier dissociation probability is not 1 and depends on some factors such as electric field and temperature. When the active layer thickness is increased to optimize the light absorption, the electric field in the blend layer decreases, which lowers down the exciton-to-free-carrier probability and makes charge collection less effective simultaneously. As a result, J_{SC} may become low, although the thickness has been optimized for better light absorption. Thus to obtain a higher J_{SC}, both the optical and the electric properties should be considered at the same time.

Some previous works (Lacic et al., 2005; Monestier et al., 2007) studied the characteristic of J_{SC}. However, they neglected the influence of exciton-to-free-carrier probability, which is important for polymer solar cells. Another study (Koster et al., 2005) considered this factor, but they neglected the optical interference effect, which is a basic property for the very thin organic film. All the above studies are based on the numerical method, and it is not easy to solve the equations and understand the direct influence of various parameters on J_{SC}. In this part, a model predicting J_{SC} is presented by using very simple analytical equations. Based on this model, the effects of optical interference on J_{SC} is investigated. Besides, the carrier lifetime is also found to be an important factor. By considering the optical interference effect and the the carrier lifetime, it is found that when the lifetimes of both electrons and holes are long enough, the exciton-to-free-carrier dissociation probability plays a very important role for a thick active layer and J_{SC} behaves wavelike with the variation of the active layer thickness; when the lifetime of one type of carrier is too short, the accumulation of charges appears near the electrode and J_{SC} increases at the initial stage and then decreases rapidly with the increase of the active layer thickness.

2.1 Theory
2.1.1 Exciton generation

The active layer in polymer solar cells absorbs the light energy when it is propagating through this layer. How much energy can be absorbed depends on the complex index of refraction $\bar{n} = n + i\kappa$ of the materials. At the position z in the organic film (Fig. 1 (a)), the time average of the energy dissipated per second for a given wavelength λ of incident light can be calculated by

$$Q(z, \lambda) = \frac{1}{2} c \varepsilon_0 \alpha_j n \left| \overline{E}(z) \right|^2 \tag{1}$$

where c is the vacuum speed of light, ε_0 the permittivity of vacuum, n the real index of refraction, α the absorption coefficient, $\alpha = 4\pi\kappa / \lambda$, and $E(z)$ the electrical optical field at point z. $Q(z, \lambda)$ have the unit of W / m^3. Assuming that every photon generates one exciton, the exciton generation rate at position z in the material is given by

$$G(z, \lambda) = \frac{Q(z, \lambda)}{h\gamma} = \frac{\lambda}{hc} Q(z, \lambda) \tag{2}$$

where h is Planck constant, and γ is the frequency of incident light. The total excitons generated by the material at position z in solar spectrum are calculated by

$$G(z) = \int_{300}^{800} G(z, \lambda) d\lambda \tag{3}$$

Here the integration is performed from 300 nm to 800 nm, which is because that beyond this range, only very weak light can be absorbed by P3HT: PCBM active layer. In inorganic solar cells, $Q(z, \lambda)$ is usually modeled by

$$Q(z, \lambda) = \alpha I_0 e^{-\alpha z} \tag{4}$$

I_0 is the incident light intensity. Here, the optical interference effect of the materials is neglected. But in polymer solar cells, the active layers are so thin compared to the wavelength that the optical interference effect cannot be neglected.

2.1.2 Optical model

In order to obtain the distribution of electromagnetic field in a multilayer structure, the optical transfer-matrix theory (TMF) is one of the most elegant methods. In this method, the light is treated as a propagating plane wave, which is transmitted and reflected on the interface. As shown in Fig. 1 (a), a polymer solar cell usually consists of a stack of several layers. Each layer can be treated to be smooth, homogenous and described by the same complex index of refraction $\overline{n} = n + i\kappa$. The optical electric field at any position in the stack is decomposed into two parts: an upstream component E^+ and a downstream component E^-, as shown in Fig. 1 (a). According to Fresnel theory, the complex reflection and transmission coefficients for a propagating plane wave along the surface normal between two adjacent layers j and k are

$$r_{jk} = \frac{\overline{n}_j - \overline{n}_k}{\overline{n}_j + \overline{n}_k} \tag{5a}$$

$$t_{jk} = \frac{2\overline{n}_j}{\overline{n}_j + \overline{n}_k} \tag{5b}$$

where r_{jk} and t_{jk} are the reflection coefficient and the transmission coefficient, $\overline{n}_j$ and $\overline{n}_k$ the complex index of refraction for layer j and layer k. So the interface matrix between the two adjacent layers is simply described as

$$I_{jk} = \frac{1}{t_{jk}} \begin{bmatrix} 1 & r_{jk} \\ r_{jk} & 1 \end{bmatrix} = \begin{bmatrix} \dfrac{\overline{n}_j + \overline{n}_k}{2\overline{n}_j} & \dfrac{\overline{n}_j - \overline{n}_k}{2\overline{n}_j} \\ \dfrac{\overline{n}_j - \overline{n}_k}{2\overline{n}_j} & \dfrac{\overline{n}_j + \overline{n}_k}{2\overline{n}_j} \end{bmatrix} \tag{6}$$

When light travels in layer j with the thickness d, the phase change can be described by the layer matrix (phase matrix)

$$L_j = \begin{bmatrix} e^{-i\beta_j} & 0 \\ 0 & e^{i\beta_j} \end{bmatrix} \tag{7}$$

where $\beta_j = 2\pi \overline{n_j} d_j / \lambda$ is phase change the wave experiences as it traverses in layer j. The optical electric fields in the substrate (subscript 0) and the final layer (subscript $m+1$) have the relationship as

$$\begin{bmatrix} \overline{E_0^+} \\ \overline{E_0^-} \end{bmatrix} = S \begin{bmatrix} \overline{E_{m+1}^+} \\ \overline{E_{m+1}^-} \end{bmatrix} = \begin{bmatrix} S_{11} & S_{12} \\ S_{12} & S_{22} \end{bmatrix} \begin{bmatrix} \overline{E_{m+1}^+} \\ \overline{E_{m+1}^-} \end{bmatrix} = \left(\prod_{v=1}^{m} I_{(v-1)v} L_v \right) \bullet I_{m(m+1)} \tag{8}$$

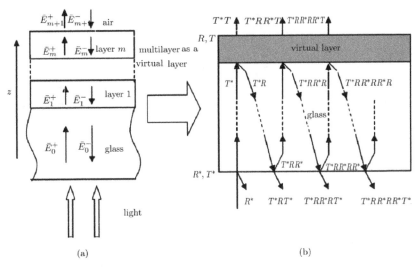

Fig. 1. Multilayer structure in a polymer solar cell. (a) the optical electric field in each layer and (b) treating the multilayer as a virtual layer.

Because in the final layer, $\overline{E_{m+1}^-}$ is 0, it can be derived that the complex reflection and transmission coefficients for the whole multilayer are:

$$r = \frac{\overline{E_0^-}}{\overline{E_0^+}} = \frac{S_{21}}{S_{11}} \tag{9a}$$

$$t = \frac{\overline{E_{m+1}^+}}{\overline{E_0^+}} = \frac{1}{S_{11}} \tag{9b}$$

In order to get the optical electric field $E_j\,(z)$ in layer j, S is divided into two parts,

$$S = S_j' L_j S_j'' \tag{10}$$

Where

$$S_j' = \left(\prod_{v=1}^{j-1} I_{(v-1)v} L_v \right) \bullet I_{j(j-1)} \tag{11a}$$

$$S_j'' = \left(\prod_{v=j+1}^{m} I_{(v-1)v} L_v \right) \bullet I_{m(m+1)} \tag{11b}$$

At the down interface in layer j, the upstream optical electric field is denoted as

$$\overline{E_j^+} = t_j^+ \bullet \overline{E_0^+} = \frac{S_{j11}''}{S_{j11}' S_{j11}'' + S_{j12}' S_{j21}'' e^{i2\beta_j}} \overline{E_0^+} \tag{12}$$

Similarly, at the up interface in layer j, the downstream optical electric field is

$$\overline{E_j^-} = t_j^- \bullet \overline{E_0^+} = \frac{S_{j21}''}{S_{j11}''} e^{i2\beta_j} \overline{E_j^+} \tag{13}$$

The optical electric field $\overline{E_j}(z)$ at any position z in layer j is the sum of upstream part $\overline{E_j^+}(z)$ and downstream part $\overline{E_j^-}(z)$

$$\overline{E_j}(z) = \overline{E_j^+}(z) + \overline{E_j^-}(z) = (t_j^+ e^{i\beta_j} + t_j^- e^{-i\beta_j}) \overline{E_0^+} \tag{14}$$

2.1.3 Light loss due to the substrate

Because the glass substrate is very thick compared to wavelength (usually 1mm>> wavelength), the optical interference effect in the substrate can be neglected. Here only the correction of the light intensity at the air/substrate and substrate/multilayer interfaces is made. As shown in Fig. 1 (b), the multilayer can be treated as a virtual layer whose complex reflection and transmission coefficients can be calculated using above equations. Then the irradiance to the multilayer is

$$I_g = T \left(\sum_{i=0}^{\infty} (R^* R)^i \right) I_0 = \frac{1 - R^*}{1 - RR^*} I_0 \tag{15}$$

I_g is described as

$$I_g = \frac{1}{2} c\varepsilon_0 n_g \left| E_0^+ \right|^2 \tag{16}$$

It can be derived that

$$\left| E_0^+ \right| = \sqrt{\frac{2(1 - R^*) * I_0}{\varepsilon_0 c n_g (1 - RR^*)}} \tag{17}$$

2.1.4 Free carrier generation

When the excitons are generated, not all of them can be dissociated into free carriers. The dissociation probability depends on the electric field and temperature. Recently, the dissociation probability has been taken into consideration in polymer solar cells [13, 16]. The geminate recombination theory, first introduced by Onsager and refined by Brau later, gives the probability of electron-hole pair dissociation,

$$P(F,T) = \frac{k_D(F)}{k_D(F) + k_X} \tag{18}$$

where k_X is the decay rate to the ground state and k_D the dissociation rate of a bound pair. Braun gives the simplified form for dissociation rate

$$k_D(F) = k_R e^{-U_B/k_B T} \left[1 + b + \frac{b^2}{3} + \cdots \right] \tag{19}$$

where a is the initial separation distance of a given electron-hole pair, U_B is electron-hole pair binding energy described as $U_B = q^2 / (4\pi\varepsilon_0\varepsilon_r a)$ and $b = q^3 F / (8\pi\varepsilon_0\varepsilon_r k_B T^2)$. T is the temperature, F the electric field and ε_r the dielectric constant of the material. In equation (19), k_R is a function of the carrier recombination. For simplification, we treat k_R as a constant. Thus, the dissociation probability P only depends on the electric field F when the temperature keeps constant.

2.1.5 J_{SC} expression equations

J_{SC} is determined by the number of carriers collected by the electrodes in the period of their lifetime τ under short circuit condition. If the active layer thickness L is shorter than the electron and hole drift lengths (which is the product of carrier mobility μ, the electric field F and the carrier lifetime τ) or in other word, the lifetimes of both types of carriers exceed their transit time (case I as in Fig 2 (a)), all generated free carriers can be collected by the electrodes. Considering the exciton-to-free-carrier dissociation probability P, J_{SC} is

$$J_{SC} = qP(F,T)GL \tag{20}$$

where G is the average exciton generation rate in the active layer.

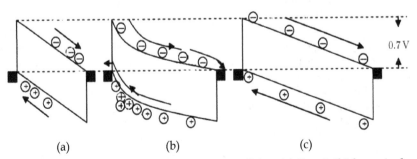

(a)   (b)   (c)

Fig. 2. Energy band diagrams under short circuit condition. (a) Case I: thickness is shorter than both drift lengths, (b) Case II: thickness is longer than hole drift length but shorter than electron drift length, c) Case III: thickness is longer than both hole and electron drift lengths.

If L is longer than drift lengths of electrons and holes, that is to say that the lifetimes of both types of carriers are smaller than their transit time, the carriers are accumulated in the active layer. At steady state, J_{SC} follows Ohm's law. Considering the exciton-to-free-carrier dissociation probability P, J_{SC} is

$$J_{SC} = qP(F,T)G(\mu_e\tau_e + \mu_h\tau_h)F = qP(F,T)G(\mu_e\tau_e + \mu_h\tau_h)V_{bi}/L \qquad (21)$$

where V_{bi} is the built-in potential which is usually determined by the difference between cathode and anode work functions. This is case III as described in Fig. 2 (c).

Between case I and case III, it is case II as described in Fig. 2 (b). In this case, L is only longer than the drift length of one type of carrier. For P3HT:PCBM based polymer solar cells, the mobilities of holes and electrons in P3HT:PCBM (1:1 by weight) layer are $2\times10^{-8}\,m^2V^{-1}s^{-1}$ and $3\times10^{-7}\,m^2V^{-1}s^{-1}$, respectively [Mihailetchi et al., 2006]. Because the hole mobility is one order lower than the electron mobility, holes are easy to accumulate in the active layer, which makes the electric field non-uniform. In order to enhance the extraction of holes, the electric field increases near the anode. On the other hand, in order to diminish the extraction of electrons, the electric field decreases near the cathode. The electric field is modified until the extraction of holes equal to the extraction of electrons. Goodman and Rose studied this case and gave an equation for the photocurrent [Goodman & Rose, 1971]. Considering the exciton-to-free-carrier dissociation probability P, J_{SC} is

$$J_{sc} = qP(F,T)GL(1+c)\frac{-c + (c^2 + 4(1-c)V\mu_h\tau_h/L^2)^{\frac{1}{2}}}{2(1-c)} \qquad (22)$$

where $c = \mu_h\tau_h/(\mu_e\tau_e)$ is the drift length ratio of holes and electrons. When $c<<1$, the equation is simplified to

$$J_{sc} = qP(F,T)G(\mu_h\tau_h)^{1/2}V^{1/2} \qquad (23)$$

2.2 Results and discussion
2.2.1 Exciton generation profile in the active layer

For the studied bulk HJ cell, the D and A materials are well blended and form the active layer. Because the D and A domains are very small, we can neglect the complex reflection and transmission at D/A interfaces, and treat the whole active layer as one homogenous material. All the optical constants (n, k) of the indium tin oxide (ITO), poly(3, 4-ethylenedioxythiophene):poly (styrene sulfonate) (PEDOT:PSS), P3HT:PCBM and the Al electrode are input into our program, and the exciton generation rate in polymer solar cells is calculated. If the interference effect is neglected, the exciton generation rate decreases with the increasing thickness of the active layer as described in equation (4) which makes the corresponding average exciton generation rate (total exciton generation rate divided by the thickness) become smaller. However, when the optical interference effect is considered, the modulation effect of average exciton generation rate with the thickness variation is very clear as shown in Fig. 3. At the initial stage, the average exciton generation rate increases with the increasing thickness of the active layer. This is because the first light peak does not appear in the active layer when the active layer is thin due to the interference effect. With the increase of the active layer, the first light peak approaches and enters the active layer such that the average generation rate becomes larger. With the further increase of the active layer, the average generation rate decreases although other light peaks enter the active layer. This is because for a thicker film, the thickness of the active layer dominates the generation rate. This evolution of exciton generation is plotted in Fig. 4 for the 500 nm wavelength.

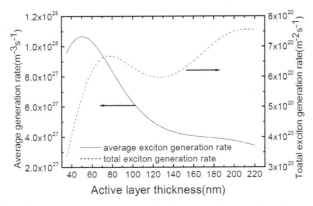

Fig. 3. The calculated exciton generation rate in the active layer when the optical interference effect is considered.

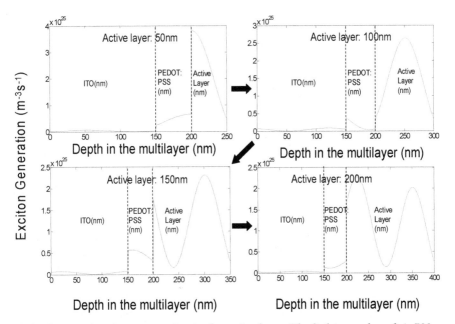

Fig. 4. Evolution of exciton generation in the active layer. The light wavelength is 500 nm. It can be seen that with the increase of the active layer thickness, the first peak enters the active layer, which makes the average exciton generation rate become large. For very thick film, although other peaks can enter the active layer, the absolute values for the peaks become small, which leads to the corresponding decrease of average exciton generation rate.

2.2.2 J_{SC} and the active layer thickness

Based on the calculated exciton generation rate, it is easy to predict J_{SC} when the drift lengths of both carriers are larger than the blend layer thickness. If all the generated excitons can be dissociated into free carriers, and then collected by the electrodes, J_{SC} should be proportional to the total exciton generation rate and behave wavelike as shown in Fig. 5 (solid line). Monestier [Monestier et al., 2007] have found this trend based on P3HT:PCBM systems. In their experiments, the active layer thickness is varied from a few tens nanometer to 215 nm. When the thickness is 70 nm, J_{SC} reaches the maximum value, and followed by a little decrease until 140 nm. When the thickness increases further, J_{SC} increases again. Unfortunately, there is obvious deviation between the prediction and the experiment results, especially in the thick film as shown in Fig. 5 (solid line). Obviously, the assumption that the exciton-to-free-carrier dissociation probability is unity is not correct. The influence of dissociation probability on J_{SC} must be considered.

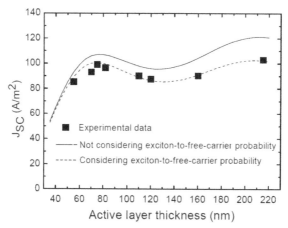

Fig. 5. Long carrier lifetime condition: the lifetimes of both carriers are always longer than their transient time. Experimental data are extracted from the work (Monestier et al. 2007).

In the previous work, Mihailetchi [Mihailetchi et al., 2006] exactly predicted photocurrent of P3HT:PCBM solar cells by assuming the same e-h separation distance (a) and decay rate (k_X). By fitting the experimental data, they obtained e-h separation distance of a=1.8 nm, room temperature bound pair decay rate of $k_X \approx 2 \times 10^4 s^{-1}$ for a 120 nm active layer, and the dissociation probability is close to 90%. We use the same data and derive the parameter $k_R = 3.9662 \times 10^8 S^{-1}$ (equation 19). The dissociation probability is calculated according to section 2.1.4. The results are shown in Fig. 6. Obviously, the exciton-to-free-carrier probability becomes lower with the increase of the active layer thickness. Using the results to correct J_{SC}, another J_{SC} curve is obtained and also shown in Fig. 5 (dash line). It can be seen that the predicted J_{SC} is exactly in accordance with the experimental results. This confirms the validity of our model. In the previous work, Monestier [Monestier et al., 2007]] modeled J_{SC} and found that the predicted J_{SC} is larger than the experimental data, especially for the thickness larger than 180 nm. They attributed this to the thickness dependence of optical constants. Here, according to our model, it is found that the deviation should come from the low exciton-to-free-carrier probability for thick active layers.

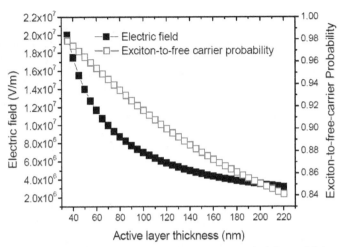

Fig. 6. Relations of electric field and exciton-to-free-carrier probability with layer thickness.

We have predicted J_{SC} precisely for the long enough carrier lifetime case. However, for polymer solar cells, the performance is sensitive to the process and experimental conditions. This may make the carrier lifetime relatively short. For P3HT:PCBM system, because the hole mobility is one order of magnitude lower than the electron mobility, holes are easy to accumulate in the active layer and limit the photocurrent. This is the case II as described in section 2.1.5. By tuning the parameters to fit the experimental data, the best fitting curve is obtained (Fig. 7) when the average hole lifetime τ is 6.2×10^{-7} s and exciton-to-free-carrier dissociation probability is unity. A short lifetime τ may imply that there are many defects.

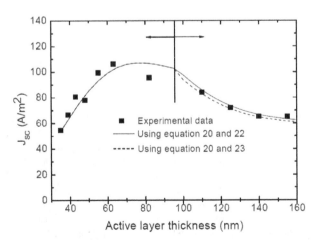

Fig. 7. Short hole carrier lifetime condition. Left arrow: hole lifetime is longer than its transient time; right arrow: hole lifetime is shorter than its transient time, and hole lifetime is 6.2×10^{-7} s , and electron lifetime is 1×10^{-6} s . Experimental data are from [Li et al., 2005].

These defects increase the exciton-to-free-carrier probability. More important, the transport process becomes the dominant limiting factor for J_{SC}, and the exciton-to-free-carrier process becomes relatively unimportant. Then it seems that the assumption of exciton-to-free-carrier probability as unity can satisfy the need of the prediction. In Fig. 7, we can see that there are two regions in the fitting curve. The left region is determined by equation (20). In this region, the lifetimes of both carriers are longer than their transient time. The solid line in the right region is determined by equation (22). In this region, hole lifetime is shorter than its transit time and electron lifetime is longer than its transit time.

If it is assumed that the drift length ratio of hole and electron is very small, then the equation (23) can be used to predict J_{SC}. As shown in Fig. 7 (dash line), it can predict J_{SC} very well, which means $c<<1$.

2.3 Summary

In this part, the exciton generation rate was calculated by taking the optical interference effect into account. Based on the calculated exciton generation rate, the dependence of J_{SC} on the active layer thickness was analyzed and compared with experimental data. Because of the optical interference effect, the total exciton generation rate does not monotonously increase with the increase of the active layer thickness, but behaves wavelike which induces the corresponding variation of J_{SC}. The carrier lifetimes also influence J_{SC} greatly. When the lifetimes of both electrons and holes are long enough, dissociation probability plays an important role for the thick active layer. J_{SC} behaves wavelike with the variation of the active layer thickness. When the hole lifetime is too short (drift length is smaller than device thickness), accumulation of charges appears near the electrode and J_{SC} increases at the initial stage and then decreases rapidly with the increase of the active layer thickness. The accordance between the predictions and the experimental results confirms the validity of the proposed model. These results give a guideline to optimize J_{SC}.

3. Effects of annealing sequence on J_{SC}

The detail of the interpenetrating network, or to say, the morphology is essentially important for the performance of polymer solar cells. In order to achieve an optimal morphology, a thermal treatment is usually utilized in the device fabrication. The thermal treatment can be carried out after and before the electrode deposition. Both the methods can greatly improve the device performance. The functions of the thermal treatment have been extensively investigated, and it has been shown that the morphology will be rearranged through the nanoscale phase separation between donor and acceptor components during the thermal treatment. By carefully optimizing the thermal treatment condition, an optimal interpenetrating network can be formed, which greatly improves the charge transport property. Besides, the thermal treatment can also effectively enhance the crystallization of P3HT, which will increase the hole mobility and the optical absorption capability. Due to the importance of the thermal treatment for P3HT:PCBM solar devices, great efforts have been devoted into the study of the thermal annealing process in the past few years. How the thermal annealing ambient, thermal annealing temperature and thermal annealing time affect the device performance has been well studied. However, only very few studies paid attention to the role of cathode in the thermal treatment. As is known, the thermal treatment can be done before and after the cathode deposition and both methods can greatly improve the device performance. The unique difference between them is whether there is cathode confinement in the thermal treatment or not. Although most of the previous studies have tended to use the cathode confinement and carry out the thermal treatment after the cathode deposition, what are the functions of the cathode confinement in the thermal treatment and how they affect the device performance are still not well studied.

In this part, the effects of cathode confinement on the performance of polymer solar cells are investigated. It is shown that a better device performance can be achieved by using the cathode confinement in the thermal treatment. The experimental analysis indicates that by capping the cathode before the thermal treatment, the Al-O-C bonds and P3HT-Al complexes are formed at the interface between the P3HT:PCBM active layer and the cathode, which leads to a better contact and thus improves the charge collection capability. More importantly, the cathode confinement in the thermal treatment greatly improves the active layer morphology. It is shown that the cathode confinement in the thermal treatment can effectively inhibit the overgrowth of the PCBM molecules, and at the same time increase the crystallization of P3HT. Thus, a better morphology is achieved, which effectively reduces the exciton loss and improves the charge transport capability. Meanwhile, the enhanced P3HT crystallites improve the absorption property of the active layer. All these effects contribute to improve the device performance.

3.1 Experimental

Fig. 8 shows the layer structure of our polymer solar cells and the chemical structures of P3HT and PCBM. All the devices were fabricated on the ITO-coated glass substrates. Briefly, after being cleaned sequentially with detergent, de-ionized water, acetone, and isopropanol in an ultrasonic bath for about 15 mins, the dried ITO glass substrates were treated with oxygen plasma for about 3 mins. Then the filtered PEDOT:PSS (Baytron P VP AI 4083) suspension (through 0.45 μm filter) was spun coated on top of the ITO surface to form a ~50 nm layer under ambient condition, and dried at 120°C in an oven for about one hour.

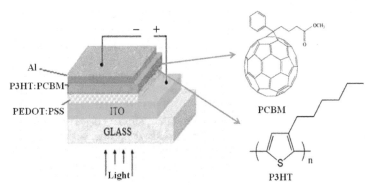

Fig. 8. Layer structure of the polymer solar cells investigated in this work.

P3HT:PCBM solution dissolved in 1,2-dichlorobenzene with a weight ratio of 1:0.8 was then spun coated on the PEDOT:PSS layer in the glove box to form a 100 nm blend layer. A 100 nm Al cathode was further thermally evaporated through a shadow mask giving an active device area of 20 mm². In order to investigate the effects of the cathode confinement on the device performance in the thermal treatment, two different types of devices were investigated: the devices without the cathode confinement in the thermal treatment (anneal the devices before the cathode deposition, pre-anneal) and the devices with the cathode confinement in the thermal treatment (anneal the devices after the cathode deposition, post-anneal). The thermal treatment was carried out by annealing the devices in the glove box at the optimized temperature of 160 °C for about 10 mins as our previous report [Zhang et al., 2008]. For reference, the devices without any thermal treatment were also fabricated.

The current-voltage (J-V) characteristics were measured by a Keithley 2400 source-measure unit under AM 1.5 solar illumination at intensity of 100 mW/cm^2 calibrated by a Thorlabs optical power meter. The XPS samples were consisted of an identical sandwiched structure: ITO coated glass/P3HT:PCBM(100 nm)/Al (3 nm). Because XPS is a surface chemical analysis technique (top 1-10 nm usually), here only a very thin metal layer is used as others [39]. The XPS spectra were measured by transferring the samples to the chamber of a Kratos AXIS HSi spectrometer at once. The operating pressure of the analysis chamber was maintained at 8x10^{-9} Torr. A 1486.71 eV monochromatic Al Kα x-ray gun source was used to achieve the Al 2p, O 1s, C 1s and S 2p spectra. Tapping mode AFM measurements were taken with a Nanoscope III A (Digital Instruments) scanning probe microscope. The samples were prepared in the same sequence as the XPS samples. The phase images and the line scanning profiles of the samples were then recorded under air operation. For both the optical absorption study and x-ray diffraction measurement, the thin films of P3HT:PCBM in the same thickness of 100 nm were spun cast on the microscope slides. The optical absorption study was recorded by a Shimadzu UV-3101 PC UV-VIS-NIR scanning spectrophotometer. The XRD measurement was carried out by the θ-2θ scan method with CuKα radiation (λ = 0.1542 nm) using a Shimadu X-Ray diffractometer.

3.2 Results and discussion

Fig. 9 shows the J-V characteristics of the devices with the same configuration of ITO/PEDOT:PSS/P3HT:PCBM/Al. For the device without any thermal treatment, it shows the solar response with J_{SC} of 5.12 mA/cm^2, V_{OC} of 0.58 V, FF of 47.63% and PCE of 1.41%. The device performance is greatly improved by the thermal treatment. However, there are obvious differences for the devices with and without the cathode confinement in the thermal treatment as shown in Fig. 9 and Table 1. For the device without the cathode confinement in the thermal treatment, it shows the performance of J_{SC}=7.50 mA/cm^2, V_{OC}=0.58 V, FF=57.13% and PCE=2.49%. However, a further performance improvement is observed for the device with the cathode confinement in the thermal treatment, which shows a better performance of J_{SC}=8.34 mA/cm^2, V_{OC}=0.60 V, FF=62.57% and PCE=3.12%. It can be seen that the cathode confinement in the thermal treatment effectively increases J_{SC} and FF, which makes the overall PCE improved by 25%. This trend was found for a series of cells. Similar results are reported recently [Kim et al, 2009] where they also observed that the device with thermal treatment after cathode deposition could show a better performance. This further confirms our experimental results.

Samples	V_{OC}	J_{SC}	FF	PCE	J_0	J_{ph}	n	R_{sh}	R_s
Without thermal treatment	0.58	5.13	47.64	1.42	2.75e-4	5.32	2.32	778.25	29.00
Without cathode confinement	0.58	7.50	57.13	2.49	4.80e-5	7.62	1.89	575.19	9.34
With cathode confinement	0.60	8.34	62.25	3.12	3.03e-5	8.40	1.88	617.28	4.43

Units of parameters, V_{OC}: V; J_{SC}, J_0 and J_{ph}: mA/cm^2; FF and PCE: %; R_{sh} and R_s: Ω cm^2.

Table 1. Summary of the Parameters Extracted from the J-V Curves Shown in Fig. 9

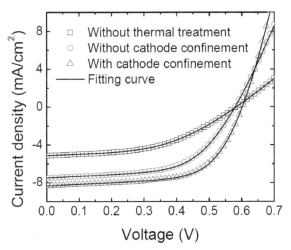

Fig. 9. *J-V* characteristics of the solar cells under the AM 1.5 illumination with the light intensity of 100 mW/cm². The devices without (circle) and with (triangle) the cathode confinement in the thermal treatment and the device without any thermal treatment (squire) are shown in the graph. Solid lines are the fitting curves according to equation (24).

In order to understand the functions of the cathode confinement in the thermal treatment, the electrical parameters need to be extracted. The *J-V* characteristics of organic solar cells can be described approximated by the Shockley equation

$$J = J_0 \left(e^{\frac{q(V - R_S J)}{n k_B T}} - 1 \right) + \frac{V - R_S J}{R_{Sh}} - J_{ph} \qquad (24)$$

Where J_0, J_{ph}, R_s, R_{sh}, q, n, k_B and T are the saturation current density, the photocurrent density, the series resistance, the shunt resistance, the electron charge, the ideality factor, the Boltzmann constant and the temperature, respectively. By fitting the Shockley equation (Fig. 9), the estimated parameters are extracted and listed in Table 1. It is shown that R_s of the device with the cathode confinement in the thermal treatment is greatly reduced compared to the device without the cathode confinement in the thermal treatment (from 9.34 Ωcm² to 4.43 Ωcm²). R_s can significantly affect the device performance and reducing the value of R_s is an efficient method to increase *PCE*. The reduced R_s by using the cathode confinement plays one main role for the significant performance improvement of polymer solar cells. R_s is directly related to the contacts between the cathode and the active layer. Thus, these contacts were addressed by the XPS measurement.

The interfacial analysis results obtained by XPS measurement are shown in Fig. 10. Each top curve and bottom curve in the Al 2p, C 1s, O 1s and S 2p core level spectra graphs are corresponding to the samples with and without cathode confinement in the thermal treatment. As shown in Fig. 10, both samples show the Al 2p spectrum peaks located at the binding energy (BE) of 74.95 eV and 74.6 eV, which are corresponding to the Al oxide and Al-O-C bond, respectively, by referring to Table 2. The Al-O-C bond is also confirmed by the peaks located at the BE of 286.2 eV in the C 1s spectrum and 531 eV in the O 1s spectrum as shown in Fig. 10. It has indicated that the Al-O-C bond is formed by the reaction of Al atoms and the carbonyl groups in PCBM and its existence will improve the contact between the polymer and the metal for both samples. However, by using the cathode confinement in thermal treatment, there is an additional shoulder peak at the BE of 76 eV in the Al 2p spectrum, which means that there forms an additional chemical bond. The additional chemical bond signal can also be seen from the S 2p spectrum. Although the typical peaks of P3HT appeared at the BE of 164.1 eV ($2p_{3/2}$) and 165.3 eV ($2p_{1/2}$) due to the spin-orbit coupling are observed for both samples in S 2p spectrum, there is an extra shoulder peak at the BE of 162.4 eV for the sample by using the cathode confinement. Considering the donation of electron density from the Al metal to the thiophene ring of P3HT, these additional peaks suggest that the interaction between P3HT and the Al metal occurs by using the cathode confinement in the thermal treatment.

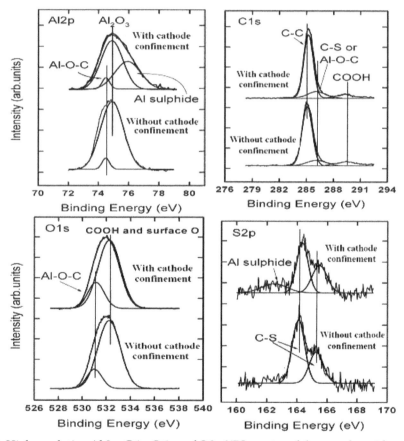

Fig. 10. High-resolution Al 2p, C 1s, O 1s and S 2p XPS spectra of the samples without (bottom curve) and with (top curve) the cathode confinement in the thermal treatment. The samples have the configuration of ITO/P3HT:PCBM (100 nm)/ Al(3 nm). By using the cathode confinement, there is an additional shoulder peak at the BE of 76 eV in the Al 2p spectrum and an additional shoulder peak at the BE of 162.4 eV in the S 2p spectrum.

Bonding states	Al 2p (eV)	C 1s (eV)	O 1s (eV)	S 2p (eV)
Al-O-C	74.6	286.2	531	
Al$_2$O$_3$	74.95		532.3	
Al-S	76			162.4
COOH		289.5		
C-C		285.1		
C-S		285.7		164.1, 165.3

Table 2. Summary of the XPS Binding Energies of Different Bonding States

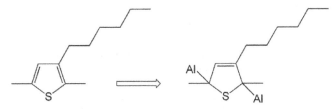

Fig. 11. The proposed molecular structure transits from P3HT to P3HT-Al complex.

Since the direct reaction between the Al atoms and the sulfur atoms is unlikely to occur because of the inherently high electron density on these sites, it is suggested that the Al atoms form bonds with the carbon atoms on the thiophene ring in the positions adjacent to the sulfur atom and form the P3HT-Al complex. One possible structure of the P3HT-Al complex is proposed in Fig. 11. The formation of the P3HT-Al complex will change the electron density of the sulfur atoms. In the P3HT-Al complex, the overall charge density of the sulfur atoms is smaller than that of the pristine P3HT. Thus the S 2p peaks located at the BE of 164.1 eV and 165.3 eV are shifted to the higher BE side at 164.3 eV and 165.5 eV, respectively, for the sample with the cathode confinement in the thermal treatment. Although the P3HT-Al complex is formed, there is only a slight energy difference (~0.1 eV shift in BE) in the C 1s spectrum for both samples as shown in Fig. 10. This is because the C 1s peak is dominated by the aliphatic carbon atoms while the Al atoms preferentially react with the carbon atoms in the conjugated system (thiophene ring of P3HT in this case). The signal arose from the interaction between P3HT and Al is too weak to affect the C 1s spectrum of the sample with cathode confinement. This explains why only very small energy difference in the C 1s spectrum is observed. The exact structure of P3HT-Al complex needs to be ascertained by further experiments.

It has been reported that Al metal can effectively transfer the electron to the conjugated polymer with the sulfide species and this feature makes it as a potential cathode for polymer electronics [Ling et al., 2002]. Another study [Reeja-Jayan et al., 2010] also has reported that Cu can react with P3HT and form sulfide-like species. The formed sulfide-like species can improve the solar cell performance. It is believed that the formation of the P3HT-Al complexes will play the same role. With the help of P3HT-Al complexes and the Al-O-C bonds, there is a better contact between the electrode and the active layer. This improved contact effectively reduces R_s and results in the improvement of the device performance.

How R_s affects the device performance is clearly shown in Fig. 12. It is shown that a large R_s will induce the decrease of FF and J_{SC}. By reducing R_s, FF and J_{SC} are increased and thus the device performance is improved. At the same time, it is also noted that although both FF and J_{SC} can be affected by R_s, their dependences on R_s are different. From Fig. 12, it can be seen that FF can be greatly adjusted by R_s when the value of R_s is just larger than 1.0 Ωcm^2. The decrease of R_s from 9.34 Ωcm^2 to 4.43 Ωcm^2 (Table 1) should be the main reason for the increase of FF from 57.13% to 62.25% (Table 1) for the sample by using the cathode confinement. However, there is no obvious change of J_{SC} observed until R_s is larger than 25 Ωcm^2 (Fig. 12). Since R_s of the two devices are relative low (9.34 Ωcm^2 and 4.43 Ωcm^2 respectively, Table 1), it seems that the decrease of R_s is not the main reason for the obvious increase of J_{SC} (from 7.50 mA/cm² to 8.34 mA/cm², Table 1) by using the cathode confinement. This conclusion is also confirmed by the extracted parameter of J_{ph}. J_{ph} is mainly determined by the properties of the active layer and only slightly depends on R_s (independent parameters in equation (24)). If the cathode confinement in the thermal treatment is only to improve the contact and reduce R_s, there should be no such obvious change of J_{ph} (from 7.62 mA/cm² to 8.40 mA/cm², Tabel I). Thus, there must be other more important factors besides R_s which lead to the obvious increase of J_{ph}. It is well known that J_{ph} is very sensitive to the device morphology and material absorption, and thus these aspects should be well addressed.

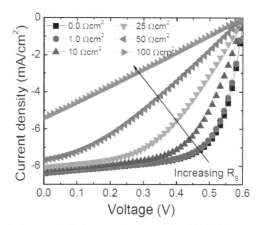

Fig. 12. Effect of R_s variation on J–V characteristics of the P3HT:PCBM solar cells according to (1). Only the value of R_s is changed while keep J_0=3.03e-5 mA/cm², J_{ph} =8.40 mA/cm², R_{sh} =617.28Ω cm² and n= 1.88. R_s greatly affects FF and thus the overall PCE. J_0, R_{sh} and n only slightly affect J_{SC} in the value range shown in Table 1.

The effects of cathode confinement on the device morphology are firstly investigated by the AFM measurement. Because the interface between the active layer and the cathode is mainly enriched by PCBM upon thermal treatment, the evolution of the surface morphology directly reflects the change of the PCBM domains. As shown in Fig. 13, it is shown that the thermal treatment effectively leads to the growth of the PCBM domains and thus increases the root mean square roughness. However, comparing to the device without the cathode confinement, there is a smoother surface morphology for the device with the cathode confinement. As shown in the AFM phase images (Fig. 13 b and c), there is a smaller island size for the sample with the cathode confinement. The profile measurements (Fig. 13 e and f) also show that the average peak-to-peak height and the width of the surface morphology are reduced by 20% and 33% by using the cathode confinement. Since surface morphology change is mainly induced by the aggregates of PCBM, the smoother surface morphology means that the cathode confinement can prevent the formation of too large underlying PCBM domains.

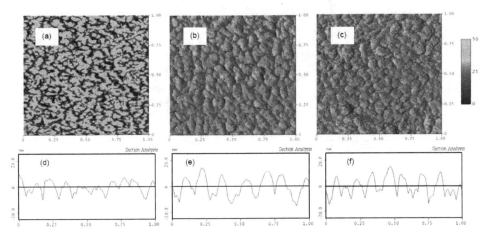

Fig. 13. Tapping-mode AFM phase images of Al covered P3HT:PCBM blend film: (a) sample without any thermal treatment and samples without (b) and with (c) the cathode confinement in the thermal treatment. Their corresponding cross sectional profiles are shown in (d) to (f) with root mean square roughness 5.5, 6.3 and 5.9 nm respectively.

It is well known that the main roles of annealing process are to induce the redistribution of PCBM and increase the crystallization of P3HT, so that the bicontinuous interpenetrating networks is achieved and meanwhile the optical absorption capability is enhanced. However, a too fast PCBM diffusion will lead to the formation of very large PCBM aggregates and thus destroy the optimal bicontinuous interpenetrating network. Besides, too large PCBM domains also reduce the interfacial contact area between P3HT and PCBM and lead to the inefficient exciton dissociation. In order to achieve a high performance, it is required to well control the PCBM domain size. It is shown here that the overgrowth of the PCBM domains in the thermal treatment is effectively inhibited by using the cathode confinement. Thus a better nanoscale morphology control is achieved. Similar metal confinement effect was also demonstrated on the organic surface by using silver cap [Peumans et al., 2003]. The improved morphology will decrease the exciton loss, facilitate the charger transport and thus increase J_{SC}.

J_{SC} is also directly related to the optical absorption of the active layer. In order to investigate the effects of cathode confinement on the optical absorption capability, the UV-Vis absorption spectra of the active layer capped with the Al electrode were measured. Because the annealed metal results in a slight variation of the light absorption, the optical spectra were obtained by subtracting the pure metal spectra. The results are shown in Fig. 14. All the samples show the typical absorption spectrum of P3HT:PCBM blend film with the absorption peak at the wavelength of 515 nm and shoulders at 550 nm and 604 nm. The thermal treatment obviously increases the optical absorption of the P3HT:PCBM film. However, there is a better optical absorption capability for the sample with the cathode confinement (e in Fig. 14) compared to the sample without the cathode confinement (d in Fig. 14). It is well known that the absorption capability of P3HT:PCBM system is directly related to the P3HT crystallites. The crystallization of P3HT was measured by XRD.

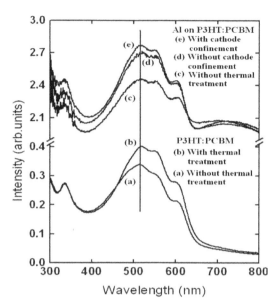

Fig. 14. Optical absorption spectra of various samples: (a) bare P3HT:PCBM blend film without thermal treatment (b) bare P3HT:PCBM blend film with thermal treatment (c) Al covered P3HT:PCBM blend film without the thermal treatment (d) Al covered P3HT:PCBM blend film with the thermal treatment done before cathode deposition (e) Al covered P3HT:PCBM blend film with the thermal treatment done after cathode deposition.

Fig. 15 shows the obtained XRD measurement results. A characteristic peak around $2\theta = 5.4°$ is observed for all the samples, which is associated with the lamella structure of thiophene rings in P3HT. Based on Bragg's law and Scherrer relation, the lattice constant (d) and the size of the polymer crystallites (L) can be determined:

$$n\lambda = 2d \sin \theta \tag{25}$$

$$L = \frac{0.9\lambda}{\Delta_{2\theta} \cos \theta} \tag{26}$$

where λ is the wavelength of the x-ray, θ the Bragg's angle, $\Delta_{2\theta}$ the smallest full width at half maximum of the peak. The extracted d and L are listed in Table 3. It is shown that all the samples show the lattice constant of 1.62 ± 0.01 nm that represents the P3HT crystallites in a-axis orientation. Thermal treatment increases the crystallization of P3HT. However, the increased magnitudes are different for the devices with and without the cathode confinement. The sample with the help of the cathode confinement in the thermal treatment shows the highest peak. As listed in Table 3, the size of the P3HT crystallites (L value of 17.7 nm) is increased by 36% by using the cathode confinement compared to without the cathode confinement (L value of 13 nm). The increased crystallite size may come from the effective inhibition of the strong PCBM diffusion by the cathode confinement. It has been shown [Swinnen et al., 2006] that a too strong diffusion of PCBM from the P3HT matrix would reduce the P3HT crystallization and optical absorption property. Because of the presence of the cathode in the thermal treatment, the PCBM diffusion is slowed down. Thus it is easier for P3HT to be crystallized. The increased P3HT crystallites will enhance the active layer optical absorption capability and increase J_{SC}.

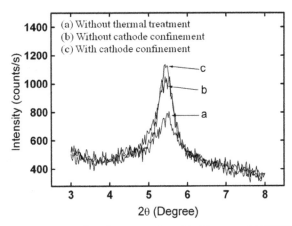

Fig. 15. X-ray diffraction spectra of various samples: (a) Al covered P3HT:PCBM blend film without the thermal treatment (b) Al covered P3HT:PCBM blend film with the thermal treatment done before cathode deposition (c) Al covered P3HT:PCBM blend film with the thermal treatment done after cathode deposition.

Samples	2θ [°]	$\Delta_{2\theta}$ [°]	h [counts/s]	L [nm]	d [nm]
Without thermal treatment	5.49	0.83	318	9.6	1.61
Without cathode confinement	5.44	0.61	596	13	1.625
With cathode confinement	5.44	0.45	617	17.7	1.625

Table 3. Summary of X-Ray Diffraction Peaks of P3HT:PCBM from Fig.15

3.3 Summary

P3HT:PCBM solar cells with the cathode confinement in the thermal treatment show better performance than the solar cells without the cathode confinement in the thermal treatment. The effects of the cathode confinement on the device performance have been investigated in this work. According to the XPS results, it is found that the Al-O-C bonds and P3HT-Al complexes are formed at the interface between the active layer and the cathode by using the cathode confinement. These chemical structures effectively reduce the contact resistance and improve the device performance. More importantly, the cathode confinement effectively improves the active layer morphology. According to the AFM, UV-Vis absorption spectra and XRD measurement results, it is found that the cathode confinement in the thermal treatment not only prevents the overgrowth of the PCBM domains, but also increases the crystallization of P3HT. With the help of cathode confinement in the thermal treatment, a better optical absorption and a more ideal bicontinuous interpenetrating networks can be obtained at the same time. This will effectively reduce the exciton loss and improve the charge transport capability. Thus an improved device performance is achieved.

4. Overall optimization of polymer solar cells

Because the active layer is very thin in compared with the incident light wavelength, the optical interference effect influences the absorption and J_{SC} as discussed in above. According to the simulated results based on the optical model, the thickness will be optimized around the first and second optical interference peaks in this part. In addition, the annealing process can efficiently improve the performance of P3HT:PCBM polymer solar cells. The performance is related to the annealing sequence, and post-annealing is more favored by the devices. Based on this conclusion, all the polymer solar cells were fabricated and post-annealed in this section. These devices were used to optimize the overall solar cell performance.

4.1 Experimental

The fabrication process is the same as above. The devices were fabricated on the ITO-coated glass substrates. After routine solvent cleaning (treated sequentially with detergent, de-ionized water, acetone, and isopropanol in an ultrasonic bath for about 15 minutes), the dried ITO glass substrates were treated with oxygen plasma for about three mins. Then the filtered PEDOT:PSS suspension was spin coated on the top of the ITO surface under ambient condition. The P3HT:PCBM solution dissolved in dichlorobenzene with a weight ratio of 1:0.8 was spin coated in the glove box. Finally, Al cathode was deposited by e-beam evaporation through a shadow mask. All the devices have same structure: ITO\PEDOT:PSS\P3HT:PCBM\Al, and only the thicknesses of the P3HT:PCBM active layers are different. The active layer thickness was controlled by changing the spin speed and solution concentration. Then different annealing temperatures are tested for the devices based on post-annealing to find the optimized conditions. The J-V characteristics were measured using a Keithley 2400 parameter analyzer in the dark and under a simulated light intensity of 100 mW/cm^2 (AM 1.5G) calibrated by an optical power meter.

4.2 Experimental results and discussion
4.2.1 Optical interference effects and active layer thickness optimization

The TMF method as discussed in section 2 is used to predict J_{SC} for the active layer thickness in a range from 50 nm to 250 nm for 1:0.8 P3HT:PCBM active layer. The results are plotted in Fig. 16. As predicted, obvious polymer solar oscillatory behavior is observed because of the very thin active layer compared with the light wavelength. When the P3HT:PCBM ratio is 1:0.8, the first and second optical interference peaks are found at the P3HT:PCBM layer thicknesses of around 85 nm and 230 nm. Both the two optical interference peaks should be used to optimize the active layer thickness.

According to the simulated results, the devices were fabricated around the first and the second optical interference peaks. The experimental results for the different active layer thicknesses are shown Fig. 17. As predicted, J_{SC} shows a periodic behavior with the variation of the active layer thickness. The J_{SC} increases from as low as 6.25 mA/cm^2 (for the device with active layer thickness, t=64 nm) to as high as 6.93 mA/cm^2 (for t=80 nm), and then decreases around the first interference peak. The same trend is observed around the second optical interference peak at a thickness of 208 nm. J_{SC} reaches a value as high as 10.37 mA/cm^2 at the second optical interference peak. The higher J_{SC} comes from the better absorption ability. It is obviously shown that the second peak can absorb more light than the first peak as shown in Fig. 18. Thus the second optical interference peak is more preferred to achieve a higher PCE. Then around this peak, the annealing conditions are investigated.

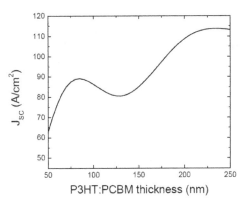

Fig. 16. J_{SC} versus P3HT:PCBM thickness, P3HT:PCBM with weight ratio of 1:0.8 and device structure of ITO/PEDOT:PSS/P3HT:PCBM/Al.

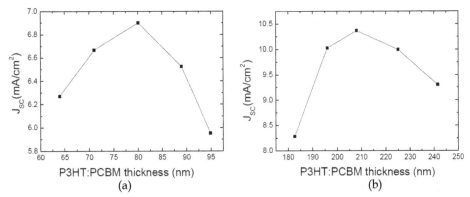

Fig. 17. Optimization of active layer thickness. (a) around the first optical interference peak, and (b) around the second optical interference peak. All devices were post-annealed at 160°C for 10 mins.

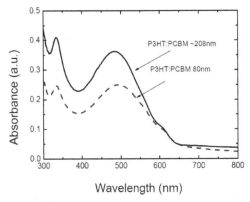

Fig. 18. UV-visible absorption spectra of P3HT:PCBM (about 80 nm thick and 208 nm thick).

4.2.2 Optimization of annealing conditions

The device performance depends greatly on annealing temperatures as clearly seen from Fig. 19. The reasons for the performance to be improved by the annealing process have been widely investigated and discussed in section 3. It is clear that for an efficient bulk HJ polymer solar cell, D and A domains must be small enough so that most of the excitons can diffuse into the D/A interfaces before they decay. At the same time, the interpenetrating transport network must be formed for the efficient charge transport. Thus, the morphology optimization is of great important. By varying the annealing condition, the morphology can be well controlled.

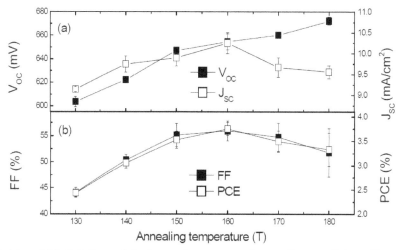

Fig. 19. (a) and (b): Relations of device performance and annealing conditions. The P3HT:PCBM layer thickness keeps constant of 208 nm.

These results were related to the better morphology as discussed in previous and also related to the increase of the charge carrier mobility. The same reason should also be responsible for our results. The highest PCE in our experiments is achieved when the annealing temperature is 160$\circ$C which is very close to the annealing temperatures reported by Ma (Ma et al., 2005). The analysis of changes in film morphology has shown that the changes in film crystallinity and aggregation within the film PCBM nanophase lead to the improved solar characteristics at this temperature. When the annealing temperature is increased, a steady enhancement of V_{OC} is observed because the e-beam evaporated Al can induce dipoles at the interface between active layer and cathode [Zhang et al., 2009]. As shown in Fig. 19, the device shows the optimized performance when it has been annealed at 160$\circ$C for 10 min.

5. Conclusion

In polymer solar cells, because of the optical interference effect, the total exciton generation rate does not increase monotonically with the increase of the active layer thickness, but behaves wave-like, which induces the corresponding variation of J_{SC}. The carrier lifetime also inffuence J_{SC} greatly. When the carrier lifetime is long enough, dissociation probability will play a very important role for a thicker active layer. J_{SC} will behave wave-like with the variation of active layer thickness. When the carrier lifetime is too short (drift length is smaller than device thickness), accumulation of charges will appear near the electrode and J_{SC} will increase at the initial stage and then decrease rapidly with the increase of active layer thickness.

The experimental studies were carried out to investigate P3HT:PCBM based HJ polymer solar cells in this chapter. It was found that the strengthened contact due to the bonding reinforcements (Al-O-C bonds and P3HT-Al complex) at the active layer/metal interface for post-annealed device improves the charge collection at the cathode side. Carrier separation can be facilitated via the improved nanoscaled morphology of the post-annealed polymer blend. The Al capping layer promotes efficient formation of the P3HT crystallites and thus enhances the light harvesting property of the polymer blend. Evidence for the latter has been derived from the improved shape of the absorption spectrum. The results underline the importance of applying the most efficient annealing sequence in order to achieve the best solar device performance.

Based on above results, the overall performance of P3HT:PCBM bulk polymer solar cells were optimized. As predicted by the TMF method, an obvious polymer solar cell oscillatory behavior of J_{SC} was observed in the experiments. The devices were optimized around the first two optical interference peaks. It was found that the optimized thicknesses are 80 nm and 208 nm. Based on the post-annealing, different annealing temperatures have been tested. The optimized annealing condition was found to be 160°C for 10 min in a nitrogen atmosphere.

6. References

Cheyns, D.; Poortmans, J.; Heremans, P.; Deibel, C.; Verlaak, S., Rand, B. P. & Genoe J. (2008). Analytical model for the open-circuit voltage and its associated resistance in organic planar heterojunction solar cells. *Physical Review B*, Vol. 77, No. 16, (April 2008), pp. 165332-1–165332-10, ISSN 1098-0121

Goodman, A. M. & Rose, A. (1971). Double Extraction of uniformly generated electron-hole pairs from insulators with nonjecting contacts. *Journal of Appllied Physics*. Vol. 42, No. 7, (June 1971). pp. 2823-2830. ISSN 0021-8979

Kim, H. J.; Park, J. H.; Lee, H. H.; Lee D. R.; & Kim, J. J. (2009). The effect of Al electrodes on the nanostructure of poly(3-hexylthiophene): Fullerene solar cell blends during thermal annealing, organic electronics, Vol. 10, No. 8, (December 2009). pp. 1505-1510. ISSN 1566-1199

Koster, L. J. A.; Smits, E. C. P.; Mihailetchi, V. D. & Blom, P. W. M. Device model for the operation of polymer/fullerene bulk heterojunction solar cells. *Physical Review B*, Vol. 72, No. 8, (August 2005) pp. 085205-1-085205-9. ISSN 1098-0121

Lacic, S. & Inganas, O. (2005). Modeling electrical transport in blend heterojunction organic solar cells. *Journal of Applied Physics*, Vol. 97, No. 12, (June 2005). pp. 124901-1-124901-7. ISSN 0021-8979

Ma, W.; Yang, C.; Gong, X.; Lee, K. & Heeger, A. J. (2005). Thermally stable, efficient polymer solar cells with nanoscale control of the interpenetrating network morphology, Advacned Functional Materials, Vol. 15, No.10, (October 2005).1617-1622. ISSN 1616-301X

Li, G.; Shrotriya, V. & Yao Y. (2005). Investigation of annealing effects and film thickness dependence of polymer solar cells based on poly(3-hexylthiophene). *Journal of Appllied Physics*.Vol. 98, No. 4 (August 2005), pp. 43704-1-43704-5. ISSN 0021-8979

Ling, Q. D.; Li, S.; Kang, E. T.; Neoh, K. G.; Liu, B. & Huang, W. (2002). Interface formation between the Al electrode and poly[2,7-(9,9-dihexylfluorene)-*co*-alt-2,5-(decylthiophene)] (PFT) investigated in situ by XPS , *Applied Surface Science*, Vol. 199, No. 1-4, (October 2002). pp. 74-82.

Monestier, F.; Simon, J. J.; Torchio, P.; Escoubas, L.; Flory, F.; Bailly, S.; Bettignies, R.; Guillerez, S. & Defranoux, C., Modeling the short-circuit current density of polymer solar cells based on P3HT:PCBM blend. *Solar Energy Materials & Solar Cells*, Vol. 91, No. 5, (March 2007). pp. 405-410. ISSN 0927-0248

Mihailetchi, V. D.; Xie, H.; Boer, B.; Koster L. J. A. & Blom, P. W. M. Charge Transport and Photocurrent Generation in Poly(3-hexylthiophene): Methanofullerene Bulk-Heterojunction Solar Cells. *Advacned Functional Materials*, Vol. 16, No. 5, (March 2006). pp. 699-708. ISSN 1616-301X

Pettersson, L. A. A.; Roman, L. S. & Inganas, O. (1999). Modeling photocurrent action spectra of photovoltaic devices based on organic thin films. *Journal of Applied Physics*, Vol. 86, No. 1, (1999). pp. 487-496. ISSN 0021-8979

Peumans, P.; Yakimov, A. & Forrest, S. R. (2003). Small molecular weight organic thin-film photodetectors and solar cells. *Journal of Applied Physics*, Vol. 93, No. 7, (April 2003). pp. 3693-3723. ISSN 0021-8979

Peumans, P.; Uchida, S. & Forrest, S. R. (2003). Efficient bulk heterojunction photovoltaic cells using small-molecular-weight organic thin films, *Nature*, Vol. 425, No. 6954, (September 2003). pp. 158-162.

Reeja-Jayan, B. & Manthiram, A. (2010). Influence of polymer–metal interface on the photovoltaic properties and long-term stabilityofnc-TiO2-P3HT hybrid solar cells,Solar Energy Materials & Solar Cells, Vol. 94, No. 5, (February 2010). pp. 907-914. ISSN 0927-0248

Swinnen, A.; Haeldermans, I.; Ven, M. V.; Haen, J. D.; Vanhoyland, G.; Aresu, S.; Olieslaeger, M. D. & Manca, J. (2006). Tuning the dimensions of C_{60}-based needlike srystals in blended thin films , *Advacned Functional Materials*, Vol. 16, pp. 760-765, 2006. ISSN 1616-301X

Zhang, C. F.; Tong, S. W.; Jiang, C. Y.; Kang, E. T.; Chan, D. S. H. & Zhu, C. X. (2008). Efficient multilayer organic solar cells using the optical interference peak, Applied Physics Letters, Vol. 93, No. 4, (August 2008). pp. 043307-1-043307-3.ISSN 0003-6951

Zhang, C. F.; Tong, S. W.; Jiang, C. Y.; Kang, E. T.; Chan, D. S. H. &Zhu, C. X. (2009). Enhancement in open circuit voltage induced by deep interface hole trapsin polymer-fullerene bulk heterojunction solar cells. Applied Physics Letters, Vol. 94, No. 10, (March 2009). pp. 103305-1-103305-3. ISSN 0003-6951

Zhang, C. F.; Hao, Y.; Tong, S. W.; Lin, Z. H.; Feng, Q; Kang, E. T. & Zhu, C. X. (2011). Effects of Cathode Confinement on the Performance of Polymer/Fullerene Photovoltaic Cells in the Thermal Treatment, *IEEE Transaction on Electron Devices*, Vol. 58, No. 3, (March 2011), pp. 835-842. ISSN 0018-9383

Flexible Photovoltaic Textiles for Smart Applications

Mukesh Kumar Singh
Uttar Pradesh Textile Technology Institute,
Souterganj, Kanpur,
India

1. Introduction

In recent years alternative renewable energies including that obtained by solar cells have attracted much attention due to exhaustion of other conventional energy resources especially fossil-based fuels. Photovoltaic energy is one of the cleanest, most applicable and promising alternative energy using limitless sun light as raw material. Even though, inorganic solar cells dominate in the world photovoltaic market, organic solar cells as the new emerging photovoltaics has explored new possibilities for different smart applications with their advanced properties including flexibility, light-weight, and graded transparency. Low cost production and easy processing of organic solar cells comparing to conventional silicon-based solar cells make them interesting and worth employing for personal use and large scale applications . Today, the smart textiles as the part of technical textiles using smart materials including photoactive materials, conductive polymers, shape memory materials, etc. are developed to mimic the nature in order to form novel materials with a variety of functionalities. The solar cell-based textiles have found its application in various novel field and promising development obtaining new features. These photovoltaic textiles have found its application in military applications, where the soldiers need electricity for the portable devices in very remote areas. The photovoltaic textile materials can be used to manufacture power wearable, mobile and stationary electronic devices to communicate, lighten, cool and heat, etc. by converting sun light into electrical energy. The photovoltaic materials can be integrated onto the textile structures especially on clothes, however, the best promising results from an efficient photovoltaic fiber has to be come which can constitute a variety of smart textile structures and related products[1].

Fossil fuels lead to the emission of CO_2 and other pollutants and consequently human health is under pressure due to adverse environmental conditions. In consequence of that renewable energy options have been explored widely in last decades[2-3].

Unprecedented characteristics of photovoltaic (PV) cells attract maximum attention in comparision of other renewable energy options which has been proved by remarkable growth in global photovoltaic market[4].

Organic solar cells made of organic electronic materials based on liquid crystals, polymers, dyes, pigment etc. attracted maximum attention of scientific and industrial community due to low weight, graded transparency, low cost, low bending rigidity and environmental friendly processing potential[5-6]. Various photovoltaic materials and devices similar to solar cells integrated with textile fabrics can harvest power by translating photon energy into electrical energy.

2. Driving forces to develop organic PV cells

Energy is the greatest technological problem of the 21st century. Energy conversion efficiency is a dominant factor to meet the increasing demand of energy worldwide. Solar energy looks easy alternative next to conventional sources, like electricity, coal and fuels. The use of solar energy can become more popular by developing photovoltaic (PV) cells of improved efficiency. The crystalline silicon PV cells are 12 % efficient with very high manufacturing cost. Thin-film cells based on CdTe, CuInS$_2$ and amorphous Si are promising, but In is expensive, Cd is toxic and amorphous Si isn't stable. A 10 % efficient cell can generate energy level equivalent to 100 W/m^2. Recently, the development of photovoltaic fibre, a great innovation in the field of photovoltaics made the technology more attractive and smart[7-12].

3. Classification of solar cells

Author has made an effort to classify the available solar cells.

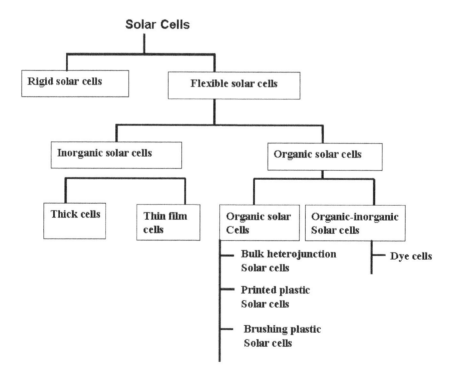

Fig. 1. Classification of Solar cells

Organic solar cells are discussed in detail in this chapter due to their higher compatibility to develop photovoltaic textiles.

4. Manufacturing of organic photovoltaic cells

Indium tin oxide (ITO) was used as a common transparent electrode in polymer-based solar cells due to its remarkable efficiency and ability of light transmission. However, it is quite expensive and generally too brittle to be used with flexible textile substrates. Therefore, highly conductive poly (3,4-ethylenedioxythiophene) doped with poly(styrene sulfonate) PEDOT:PSS, carbon nano-tube or metal layers are used to substitute ITO electrode. This can be a promising way to develop PV textiles for smart application due to its low cost and easy application features for future photovoltaic textile applications. A typical sequence of photovoltaic textiles manufacturing is exhibited in Fig. 2.

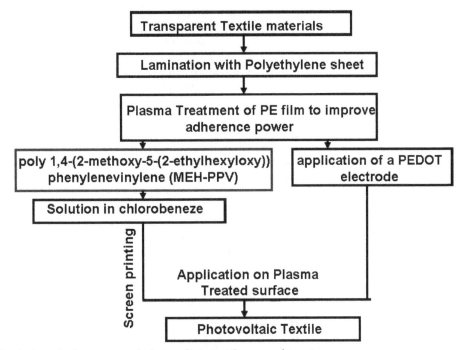

Fig. 2. A typical sequence of photovoltaic textiles manufacturing

A group of scientists has demonstrated the fabrication of an organic photovoltaic device with improved power conversion efficiency by reducing lateral contribution of series resistance between subcells through active area partitioning by introducing a patterned structure of insulating partitioning walls inside the device. Thus, the method of the present invention can be effectively used in the fabrication and development of a next-generation large area organic thin layer photovoltaic cell device[13].

The manufacturing of organic photovoltaic (PV) cells can be possible at reasonable cost by two techniques:

4.1 Roll-to-roll coating technique

A continuous roll-to-roll nanoimprint lithography (R2RNIL) technique can provide a solution for high-speed large-area nanoscale patterning with greatly improved throughput. In a typical process, four inch wide area was printed by continuous imprinting of nanogratings by using a newly developed apparatus capable of roll-to- roll imprinting (R2RNIL) on flexible web base. The 300 nm line width grating patterns are continuously transferred on flexible plastic substrate with greatly enhanced throughput by roll-to- roll coating technique.

European Union has launched an European research project "HIFLEX" under the collaboration with Energy research Centre of the Netherland (ECN) to commercialize the roll to roll technique. Highly flexible Organic Photovoltaics (OPV) modules will allow the cost-effective production of large-area optical photovoltaic (OPV) modules with commercially viable Roll-to-Roll compatible printing and coating techniques.

Coatema, Germany with Renewable Technologies and Konarka Technologies has started a joint project to manufacture commercial coating machine. Coatema, Germany alongwith US Company Solar Integrated Technologies (SIT) has developed a process of hot-melt lamination of flexible photovoltaic films by continuous roll-to-roll technique[14]. Roll-to-roll (R2R) processing technology is still in neonatal stage. The novel innovative aspect of R2R technology is related to the roll to roll deposition of thin films on textile surfaces at very high speed to make photovoltaic process cost effective. This technique is able to produce direct pattern of the materials[15, 16].

4.2 Thin -film deposition techniques

Various companies of the world have claimed the manufacturing of various photovoltaic thin films of amorphous silicon (a-Si), copper indium selenide (CIGS), cadmium telluride (CdTe) and dye-sensitized solar cell (DSSC) successfully. Thin film photovoltaics became cost effective after the invention of highly efficient deposition techniques. These deposition techniques offer more engineering flexibilities to increase cell efficiencies, reflectance and dielectric strength, as well as act as a barrier to ensure a long life of the thin film photovoltaics and create high vapour barrier to save the chemistry of these types of photovoltaics[17-18].

A fibre shaped organic photovoltaic cell was produced by utilizing concentric thin layer of small molecular organic compounds as shown in Fig 3.

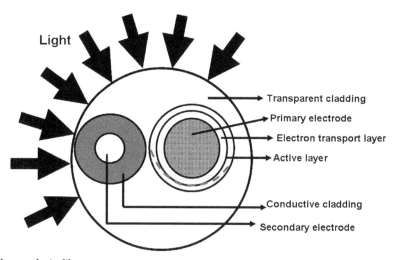

Fig. 3. Photovoltaic fibre

Thin metal electrode are exhibited 0.5% efficiency of solar power conversion to electricity which is lower than 0.76% that of the planner control device of fibre shape organic PV cells. Results are encouraged to the researchers to explore the possibility of weaving these fibres into fabric form.

4.2.1 Dye-sensitized photovoltaics

An exhaustive research on photovoltaic fibres based on dye-sensitized TiO_2-coated Ti fibers has opened up various gateways for novel PV applications of textiles. The cohesion and adhesion of the TiO_2 layer are identified as crucial factors in maintaining PV efficiency after weaving operation. By proper control of tension on warp and weft fibres, high PV efficiency of woven fabrics is feasible.

The deposition of thin porous films of ZnO on metalized textiles or textile-compatible metal wires by template assisted electro-deposition technique is possible. A sensitizer was adsorbed and the performance as photoelectrodes in dye-sensitized photovoltaic cells was investigated. The thermal instability of textiles restricts its use as photovoltaic material because process temperatures are needed to keep below 150°C. Therefore, the electro-deposition of semiconductor films from low-temperature aqueous solutions has become a most reliable technique to develop textile based photovoltaics. Among low-temperature solution based photovoltaic technologies; dye sensitized solar cell technology appears most feasible. If textile materials are behaved as active textiles, the maximum electrode distance in the range of 100 µm has to be considered. Loewenstein et al., (2008) and Lincot et al., (1998) have used   Ag coated polyamide threads and fibers to deposit porous ZnO as semiconductor material . The crystalline ZnO films were prepared in a cathodic electrodeposition reaction induced by oxygen reduction in an aqueous electrolyte in presence of Zn^{2+} and eosinY as structure-directing agent[19-20].

Bedeloglu et al., (2009)[21] were used nontransparent non-conductive flexible polypropylene (PP) tapes as substrate without use of ITO layer. PP tapes were gently cleaned in methanol, isopropanol, and distilled water respectively and then dried in presence of nitrogen. 100nm thick Ag layer was deposited by thermal evaporation technique. In next step, a thin layer of poly(3,4-ethylenedioxythiophene) doped : poly(styrene sulfonate) PEDOT: PSS mixture solution was dip coated on PP tapes. Subsequently, poly [2-methoxy-5-(3, 7-dimethyloctyloxy)-1-4-phenylene   vinylene]   and   1-(3-methoxycarbonyl)-propyl-1-phenyl(6,6)C61, MDMO: PPV: PCBM or poly(3-hexylthiophene) and 1-(3-methoxycarbonyl)-propyl-1-phenyl(6,6)C61, P3HT: PCBM blend were dip coated onto PP tapes. Finally, a thin layer of LiF (7nm) and Al (10nm) were deposited by thermal evaporation technique.

The enhanced conductivity will always useful to improve the photovoltaic potential of poly(3,4-ethylene dioxythiophene):poly(styrene sulfonate) (PEDOT:PSS). Photovoltaic scientific community found that the conductivity of poly(3,4-ethylene dioxythiophene): poly(styrene sulfonate) (PEDOT:PSS), film is enhanced by over 100-folds if a liquid or solid organic compound, such as methyl sulfoxide (DMSO), N,Ndimethylformamide (DMF), glycerol, or sorbitol, is added to the PEDOT:PSS aqueous solution. The conductivity enhancement is strongly dependent on the chemical structure of the organic compounds. The aqueous PEDOT: PSS can be easily converted into film form on various substrates by conventional solution processing techniques and these films have excellent thermal stability and high transparency in the visible range[22-25].

Some organic solvents such as ethylene glycol (EG), 2-nitroethanol, methyl sulfoxide or 1-methyl-2-pyrrolidinone are tried to enhance the conductivity of PEDOT: PSS. The PEDOT:

PSS film which is soluble in water becomes insoluble after treatment with EG. Raman spectroscopy indicates that interchain interaction increases in EG treated PEDOT: PSS by conformational changes of the PEDOT chains, which change from a coil to linear or expanded-coil structure. The electron spin resonance (ESR) was also used to confirm the increased interchain interaction and conformation changes as a function of temperature. It was found that EG treatment of PEDOT: PSS lowers the energy barrier for charge among the PEDOT chains, lowers the polaron concentration in the PEDOT: PSS film by w 50%, and increases the electrochemical activity of the PEDOT: PSS film in NaCl aqueous solution by w100%. Atomic force microscopy (AFM) and contact angle measurements were used to confirm the change in surface morphology of the PEDOT: PSS film. The presence of organic compounds was helpful to increase the conductivity which was strongly dependent on the chemical structure of the organic compounds, and observed only with organic compound with two or more polar groups. Experimental data were enough to make a statement that the conductivity enhancement is due to the conformational change of the PEDOT chains and the driving force is the interaction between the dipoles of the organic compound and dipoles on the PEDOT chains[26].

Thin film PV structure offers following advantages [27-29]:

- Photovoltaic thin film structures are more efficient in comparison to their planar counterparts.
- Photovoltaic thin films offer increased surface area which is favourable for light trapping due to a reduction in specular reflectance but increased internal scattering, leading to increased optical path lengths for photon absorption.
- In Photovoltaic thin film structures, transport lengths for photoexcited carriers in the absorber are reduced and so electrons and holes do not need to travel over large distances before separation and collection.

4.2.2 Thin -film deposition technique

The thin film deposition of photovoltaic materials takes place by electron beam, resistance heating and sputtering techniques. These technologies differ from each other in terms of degree of sophistication and quality of film produced. A resistance-heated evaporation technology is relatively simple and inexpensive, but the material capacity is very small which restricts its use for commercial production line. Sputtering technique can be used to deposit on large areas and complex surfaces. Electron beam evaporation is the most versatile technique of vacuum evaporation and deposition of pure elements, including most metals, numerous alloys and compounds. The electron beam technology has an edge over its counterparts due to following merits of this technology:

- precise control at low or high deposition rates is possible
- possibilities of co-deposition and sequential deposition systems are available
- uniform low temperature deposition is possible
- excellent material utilization is possible
- higher evaporation rates are possible
- freedom from contamination is possible
- precise film composition and cooler substrate temperatures can be maintained

4.2.2.1 e-Vap® thin film deposition technology

Various frames of different electron beam sizes are offered by e-Vap® which are able to produce small research specimen to achieve commercial coating requirement with crucible capacities from 2cc to 400cc. e-Vap® 100 miniature evaporation systems is a precise wire-fed electron beam source designed specifically for depositing monolayer thin films in ultrahigh vacuum environments capable to deposit metals at atomic level. e-Vap® 3000 and Caburn-MDC e-Vap® are other electron beam evaporation system of different capacity for a wide range of applications[30]. Various companies are working in the field of thin film photovoltaics as shown in Table 1.

Major companies	Technology	Status of manufacturing
Siemens Solar Industries (SSI), Global Solar	Copper Indium Diselenide	Initial Small Quantity Manufacture under 100 kW at SSI
First Solar, BP Solar, Matsushita	First Solar, BP Solar, Matsushita	First Solar Production under 1 MW, Others Lower
Solarex, United Solar, Canon, others	Amorphous Silicon	Commercial Production under 10 MW at Several Plants

Table 1. Photovoltaic thin film manufacturing

4.3 Printing of plastic solar cells
Organic semiconductor based solar cells can be integrated fast with textile substrates and molecular heterojunction cells can be printed using inkjet printing efficiently. This technology has opened new routes to produce organic solar cells. Credit of invention of printed solar cells goes to Konarka Technologies[31] for successful demonstration of manufacturing of solar cells by inkjet printing as shown in Fig.4 .

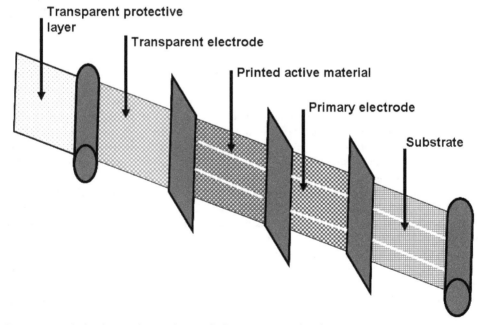

Fig. 4. Konarka's plastic photovoltaic cells by printing technology

The inkjet printing technology enables manufacturing of solar cells with multiple colors and patterns for lower power requirement products, like indoor or sensor applications. A mixture of high and low boiling solvents, (68% orthodichlorobenzene and 32% 1,3,5-trimethylbenzene), is found suitable for the production of inkjet printed organic solar cells with power conversion efficiency upto 3%. During the drying process and subsequent annealing, the suggested oDCB–mesitylene solvent mixture leads to an optimum phase separation network of the polymer donor and fullerene acceptor and therefore strongly enhances the performance. During drying and subsequent heat-setting process, the recommended ortho-dichlorobenzene (oDCB)-mesitylene solvent mixture leads to an optimum phase separation of polymer donor and fullerene acceptor as suggested by Pagliaro et al., (2008)[32]. Solvents formulation and temperature of printing table are two prime parameters to control the spreading and wetting of liquid on substrate surface. Fig.4 shows a schematic representation of organic film formation by inkjet printing.

In a typical case, the photoactive formulation is formed by blending poly(3-hexylthiophene) (P3HT) with fullerene [6,6]-phenyl C61 butyric acid methyl ester (PCBM) in a tetralene and oDCB–mesitylene solvent mixture. A uniform film and reliable printing with respect to the spreading and film formation was performed by keeping the inkjet platen temperature 40°C. The combination of higher/lower boiling solvent mixture, oDCB–mesitylene, offers following advantages:

a. oDCB with b.p.¼180°C can be used to prevent nozzle clogging and provide a reliable jetting of the printhead

b. the second component, mesitylene, with lower boiling point of 165°C of the solvent mixture, with a lower surface tension, is used to achieve optimum wetting and spreading of the solution on the substrate. It has a higher vapor pressure of 1.86mm Hg at 20°C and a lower boiling point of 165°C compared to oDCB and tetralene. It increases the drying rate of the solvent mixture, which is a critical parameter to decide the morphology of PV prints.

According to Hoth et al., (2007) for an efficient bulk heterojunction solar cell, precise control of the morphology is essential. The active layer deposition tool strategy decides the morphology. It was evident from AFM study of the inkjet printed active layers that the P3HT–PCBM blend films show significant difference in the grain size and surface roughness. The roughness of active layer surface affects the performance of the inkjet printed photovoltaic device. The credit of commercialization of power plastic cells (PPC) goes to Konarka alongwith a German firm Leonhard Kurz by opting simple, energy efficient, environmentally friendly, replicable and scalable process. The semiconducting conjugated polymers to make the photosensitive layers of the cell are created in batches of several liters each. Finally fluffy powder is formed and manufacturers combine it with standard industrial solvents to create an ink or coatable liquid. This coatable liquid is fed in reservoir of inkjet print head. Specific types of pumps are used to exert continuous pressure to maintain constant through put rate from orifice inkjet printhead throughout the printing process. Inkjet head has facility to move in different directions which helps to create various printing patterns of semiconducting polymer liquid on textile substrate layer by layer as shown in Fig.5. These layers are considerably thin. During deposition of semiconducting polymer cleanliness is very important and whole printing process is carried out in a clean room[31].

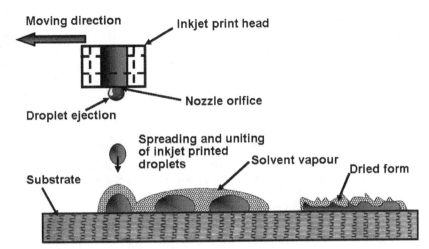

Fig. 5. Inkjet printing of photovoltaic cells

4.3.1 Advantages of inkjet printing

Inkjet printing of organic photovoltaic materials offers following advantages:

- Inkjet printing is a commonly used technique for controlled deposition of solutions of functional materials in specific locations on a substrate and can provide easy and fast deposition of polymer films over a large area.
- Organic solar cells can be processed with printing technologies with little or no loss compared to clean room, semiconductor technologies such as spin coating.
- Inkjet printing could become a smart tool to manufacturer solar cells with multiple colors and patterns for lower power requirement products, like indoor or sensor applications.
- Inkjet printing technique is considered very promising because the polymer devices can be fabricated very easily because of the compatibility with various substrates and it does not require additional patterning.

5. Significance of low bandgap polymers

The part of visible sunlight is lost by absorption in specific regions of the spectrum when passes through the atmosphere. The amount of loss depends on the air mass. Under ideal conditions, the available photons for the conversion to the electrons can be represented by the solar spectrum in photon flux as a function of wavelength. It is evident that photons must be harvest at longer wavelength but at longer wavelengths the energy of the charge carriers remains lower which restricts the voltage difference that the device can produce. Hence designing of optimum bandgap is essential. However, the practical efficiencies differ from theoretically predicted values. These above considerations are based on the fact that the low band gap polymers have the possibility to improve the efficiency of OPVs due to a better overlap with the solar spectrum. Hence to achieve maximum power generation in photovoltaic device low band gap materials are required. Majority of Photovoltaic devices are unable to convert light energy below 350–400 nm wavelengths efficiently into electrical energy because of poor absorption in the substrate and front electrodes. Although, this part of light spectrum contains very little intensity and consequently do not have a major contribution and i.e only 1.4% to the total possible current. It is evident from above discussion that to increase the current realization λ_{max} have to increase from 650 to 1000nm, in turn decreasing the band gap. Poly (3-hexylthiophene) is a typical example of low badgap polymeric material has a band gap of 650nm (1.9 eV) which can harvest up to 22.4% of the available photons. Hence, it is necessary to fabricate the polymers having broad absorption to achieve increase in the efficiency of the solar cell[33-34].

6. Different techniques to manufacture photovoltaic textiles

Photovoltaics and textiles are two different areas where a successful collaboration brings very smart results. The integration of these two different sectors can be possible by adopting following techniques.

There are two basic techniques to manufacture PV textiles.

6.1 PV textiles by PV fibres

This technique is based on the development of photovoltaic fibres using Si-based /organic semiconducting coating or incorporation of dye-sensitized cells (DSC). Availability of PV fibre offer more freedom in the selection of structure for various type of applications[35-39].

The development of photovoltaic fibres offers advantages to manufacture large area active surfaces and higher flexibility to weave or knit etc[40].

Although, the problem of manufacturing textile structure by using dye-sensitized cells (DSC-PV) fibres into textile structure is still alive and require a optimization with respect to textile manufacturing operations. In a typical research work the working electrode of DSC-PV fibre is prepared by coating Ti wire with a porous layer of TiO_2. This working electrode is embedded in an electrolyte with titanium counter electrode. The composite structure is coated with a transparent cladding to ensure protection and structural integrity. The electrons from dye molecules are excited by photo energy and penetrated into the conduction band of TiO_2 and move to the counter electrode through external circuit and regenerate the electrolyte by happening of redox reaction. Ultimately the electrolyte regenerates the dye by means of reduction reaction.

The performance of DSC fibre is majorily depends on the grade of TiO_2 coating and its integration to Ti substrate. The integrity of Ti with TiO_2 will depend on the surface cleanliness and roughness of Ti, affinity between Ti and TiO_2 and other defects. The deposition of TiO_2 dye on Ti wire surface is performed by strategy as shown in Fig.6. The integrity of coating on Ti substrate is tested by using peel test, tensile test, four point bending test and scratch test. The amount of discontinuities is measured by optional microscopy and SEM[41-47].

The photovoltaic potential of dye-sensitized solar cells (DSSC) of Poly(vinyl alcohol) (PVA) was improved by spun it into nanofibers by electrospinning technique using PVA solution containing silver nitrate ($AgNO_3$). The silver nanoparticles were generated in electrospun PVA nanofibers after irradiation with UV light of 310~380 nm wavelength. Electrospun PVA/Ag nanofibers have exhibited I_{sc} , FF, V_{oc} , and η showed the values of 11.9~12.5 mA/cm^2, 0.55~10.59, 0.70~10.71 V, and 4.73~14.99%, respectively. When the silver was loaded upto 1% as dope additives in PVA solution, the resultant electrospun PVA/Ag nanofibres exhibited power conversion efficiency 4.99%, which is higher than that of dye sensitized solar cells (DSSC) using electrospun PVA nanofibers without Ag nanoparticles[48].

Ramier et al., (2008), concluded that the feasibility of producing textile structure from DSC-PV fibre is quite good. The deposition of TiO$_2$ on flexible fibre is expected to be quite fruiteful in order to maintain the structural integrity without comparing with PV performance[49].

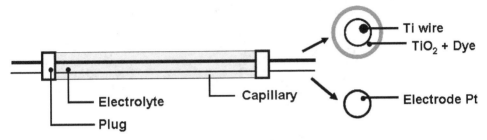

Fig. 6. A model DSC Photovoltaic fibre by surface deposition

Fibre based organic PV devices inroads their applications in electronics, lighting, sensing and thermoelectric harvesting. By successful patch up between commodity fibre and photovoltaic concept, a very useful and cost effective way of power harvesting is matured[50-52].

Coner et al.,[53] have developed a photovoltaic fibre by deposition of small Molecular weight organic compound in the form of concentric layer on long fibres. They manufactured the OPV fibre by vacuum thermal evaporation (VTE) of concentric thin films upto 0.48 mm thickness on polyamide coated silica fibre. Different control devices are based on OPV cells containing identical layer structures deposited on polyimide substrates. The OPV based fibre cells were defined by the shape of the substrate and 1 mm long cathodes. All fibre surfaces were cleaned well prior to deposition. Lastly, they concluded that performance of OPV fibre cells from ITO is inferior in terms of changes in illumination angle, enabling the optical photovoltaic (OPV) fibre containing devices to outperform its planar analog under favourable operating conditions. Light emitting devices are designed in such a way that becomes friendly to weave it. The light trapping on fibre surface can be improved by using external dielectric coating which is coupled with protective coating to enhance its service time. Successful PV fibre can be manufactured by opting appropriate material with more improved fabrication potential[54].

Dye sensitized solar cells (DSC) are low cost, applicable in wide range of application and simple to manufacture. These merits of dye-sensitized PV fibre makes it a potential alternative to the conventional silicon and thin film PV devices[55].

DSC works on the principle of optoelectronically active cladding on an optical fibre. This group was manufactured two type of PV fibre using polymethylmethacrylate (PMMA) baltronic quality diameter 1.3 to 2.0mm and photonium quality glass fibre with diameter 1.0 to 1.5 mm. Both virgin fibre were made electronically conductive by deposition of 130nm thick layer of ZnO:Al by atomic layer deposition technique with the help of P400 equipment. The high surface area photoelectric film for DSC was prepared in two steps. In first step TiO$_2$ in the form of solution or paste having TiO$_2$ nanoparticle is deposited on electronically conductive surface. In the second step dry layer of TiO$_2$ is sintered at 450-500°C for 30 minute to ensure proper adhesion to the fibre surface. PMMA fibre is suitable to survive upto 85°C. Hence mechanical compression is alternate technique to ensure the fixation.

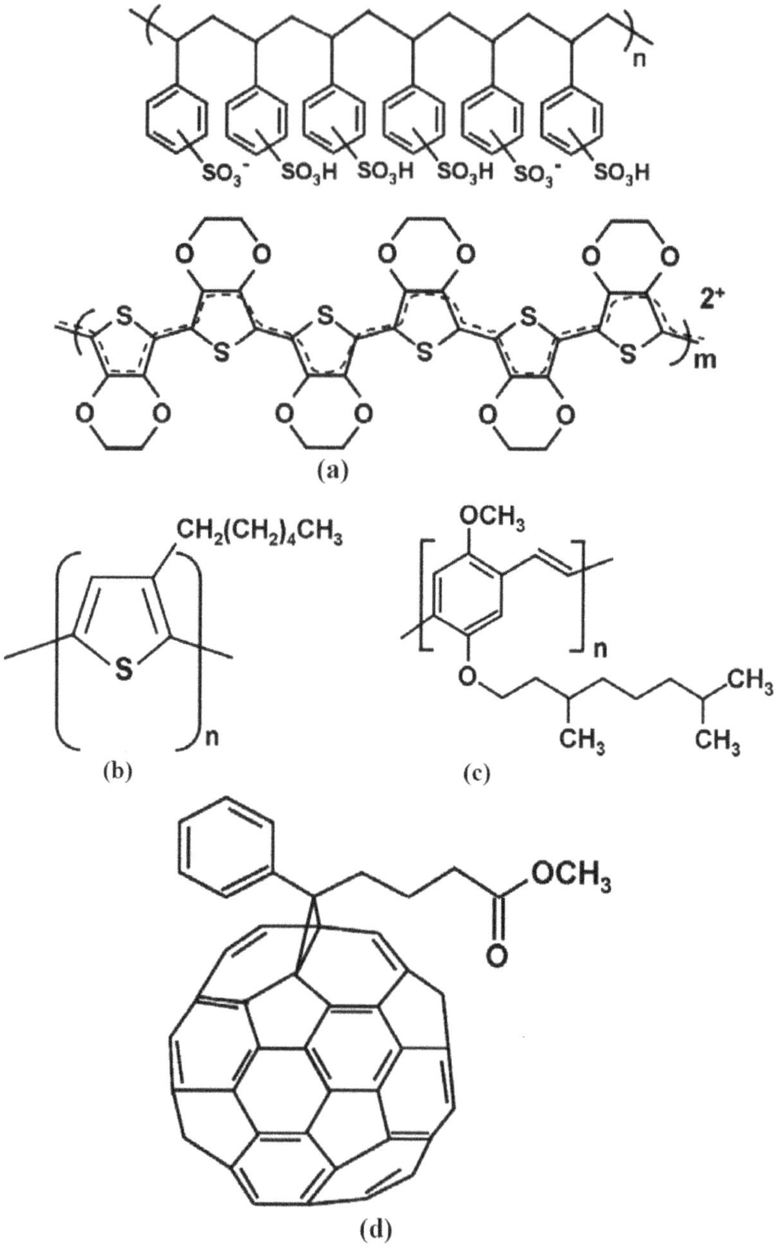

Fig. 7. Chemical configurations of (a) PEDOT:PSS (b) P3HT (c) MDMO-PPV (d) PCBM

Glass fibre is capable to withstand with sintering temperatures which inroads the possibilities of preparation of porous photoelectrodes on them. Commercially available TiO_2 paste was diluted to achieve appropriate viscosity with tarpin oil to make suitable for dip-coating. TiO_2

film was formed by dip coating and dried at room temperature upto 30 min before proper sintering between 475 to 500°C. Appropriately sintered fibres were then immersed in dye solution consisting of 0.32 ml of the cis-bis(isothiocyanate)bis(2,2-bipyridyl-4-4' dicarboxylato)-ruthenium(II) bis-tetrabutyl ammonium, Solaronix SA with trade name N719 dye in absolute ethanol for 48h. The dyeing of nonporous Polymethylmethacrylate (PMMA) fibre coated with nonporous TiO_2 layer was performed in the same dye bath. After complete senetization the excess dye was rinsed away with ethanol. A electrolyte solution was prepared with 0.5 M 4-tert butylpyridine and 0.5 M Lil, 0.05MI2 in 3 methoxypropionitrile (MePRN) with 5 wt% polyvinylidene fluoride-hexafluoro-propylene(PVDE-HFP) added as gelatinizing agent as used by Wang et al., (2004)[56].

Finally gelatinized iodine electrolyte was added next with dip coating from hot solution. Lastly the carbon based counter electrode was coated by means of a gel prepared by exhaustive grinding of 1.4g graphite powder and 0.49 grade carbon black simultaneously.

6.1.1 Manufacturing of photovoltaic fibres as per Bedeloglu et al.,[57-58] method

Bedeloglu et al., have used nontransparent PP as substrate. The PP tape was washed using methanol, isopropanol and water and then dried in N2 atmosphere. Thermal evaporation technique was used to deposit 100nm thick Ag contact layer on PP substrate. A filtered solution of PEDOT: PSS, chemical structure shown in Fig.7, in 5% dimethyl sulfoxide (DMSO) and stirred with 0.1% Triton X-100 to increase the thermal conductivity and cohesiveness properties. Stirred mixture of PEDOT: PSS were deposited on cleaned PP tapes at a thickness of 200 nm by dip coating technique. A blend of P3HT as conjugated polymer and PCBM was stirred upto 24h in chlorobenzene and then dip coated with thickness of 200 nm on top of PEDOT: PSS layer. Finishing of all PV structures was completed in vacuum chambers. An aluminium contact layer of approximately 3 nm thickness was deposited followed by 7nm thick Ag layer as anode. The purpose of Al layer was to avoid short circuiting between Ag and PEDOT: PSS films. The resultant photovoltaic fibre is shown in Fig.8.

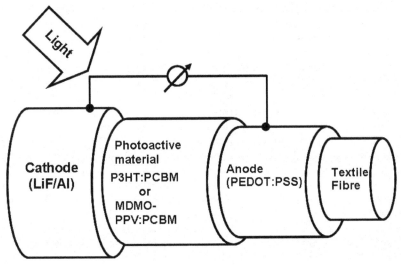

Fig. 8. Schematic representation of Photovoltaic fibre

A group of Turkish scientists has standardized a photovoltaic fibre manufacturing process. A monofilament supply cope is used to supply the basic filament for PV fibre manufacturing. This monofilament is cleaned in a bath by methanol solution and then further clean up by isopropanol solution in second bath. The cleaned fibre surface is washed with distill water followed by drying with dry nitrogen flow. The fibre is immersed in fourth bath containing mixture of PEDOT: PSS followed by oven drying at 50°C for 3 hr. The coated fibres are further immersed in another subsequent bath containing photoactive material solution and then dried out at 50°C for 15 min in oven. After drying, deposition of metal electrode takes place on fibre surface followed by deposition of anti-reflective materials by appropriate deposition technique. Finally a protective layer is laminated on fibre surface. In consequence of this process, a photovoltaic fibre is manufactured and become ready for power harvesting.

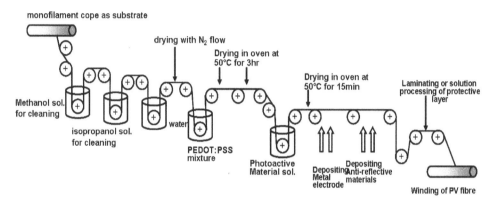

Fig. 9. Manufacturing process of Photovoltaic fibre as suggested by Bedeloglu et al.[21]

6.2 PV textiles by patching (attachment of PV cells to existing textile structures)

Under this technology a solar cell is manufactured separately and then patched onto textile structure by different ways. Thin film solar cells with adequate flexibility can be patched successfully on textile surfaces to impart PV effect. This technique is most appropriate to easily develop both small and large area structures with low cost, and light weight.

In a typical approach thin layers of polymer photovoltaics are laminated onto a textile substrate followed by plasma treatment and coating of PEDOT electrode. Subsequent screen printing of the active material and evaporation of the ultimate electrode finished the polymer photovoltaic that is fully integrated into the textile material[59].

Poly 1,4-(2-methoxy-5-(2-ethylhexyloxy))phenylenevinylene (MEH-PPV) was synthesized by polymerization of α,α'-dibromo-2-methoxy-5-(2-ethylhexyloxy) xylene as described by Neef and Ferraris. The purified MEH-PPV was characterized and found the Mw of 249,000 g mol[-1] , polydispersity of 5.46 and a peak molecular weight Mp of 157,500 g mol[-1]. A polyethylene terephthalate (PET) film of thickness 200 μm covered with 50 Ω^2 ITO layer was iched by 20% HCL and 5%HNO$_3$. A 250 μm thick polyethylene film was thermally laminated on polyethylene terephthalate (PET) surface. Both PET and PE surfaces was plasma treated using a 350-1 low power 3-phase AC plasma system as prescribed by Jensen and Glejbol (2003) in order to obtain optimum etching to get appropriate adherence with textile substrates[60-61].

Plasma treatment was followed by application of PEDOT electrode deformation by spray painting of an inhibited mixture of Iron (III) Tosylate and 3,4-Ethylene-dioxythiopene (EDT) through an aluminum mask. As the temperature reaches 50°C inhibitors are started to evaporate and polymerization of EDT initiated. After completing the polymerization, the PEDOT-coated textile was washed in cold water twice to washout excess tosylate, Iron (II) and inhibitor residuals and finally annealed in air at 50°C for 2 hour.

The materials that can generate electricity by photon conversion are loaded with fibres, yarns and textile structures inherently to offer PV effect.

Chlorobenzene was used to get the solution of MEH-PPV of 3.5gL^{-1}. The solution was filtered through a filter of 2.7µm filter and coated on polyethylene terephthalate- Indium tin oxide (PET-ITO) substrate by doctor blade coating system. Consequently a homogeneous red colour film is coated on PET-ITO substrate. Cholorobenezene solution of 25gL^{-1} was used to get printed pattern of polymer by means of screen-printing. The masks were prepared by using threads of 27 micron diameter with 140, 180,200 and 220 fibres cm^{-1} mesh. Finally, the printed substrates were dried in absence of sunlight. Photovoltaic device was prepared by following manner. The PET-ITO-MEHPPV substrates were incorporated behind a mask and a layer of C60 and aluminium was deposited by thermal evaporation technique. After completing the evaporation the device was laminated on 100 micron PET substrate on electrode side. The screen printed photovoltaic textile was mounted on large evaporator and kept the distance between thermal source and substrate 65 cm. The PV textile holder was rotated at 30 rpm. The aluminium electrodes were prepared in the form of 300 nm thick layers. The evaporation chamber was filled with dry nitrogen after completing the 30 min cooling. Electrodes were integrated with textile substrate by using silver epoxy[62].

7. Characterization of photovoltaic textiles

Characterization of various photovoltaic textiles is essential to prove its performance before send to the market. Various characterization techniques collectively ensures the perfect achievement of the targets to manufacture the desired product.

7.1 Thickness and morphology of photovoltaic textiles

Scanning electron microscope is used to investigate the thickness and morphology of various donor, acceptor layers. Scanning electron microscopes from LEO Supra 35 and others can be used to measure the existence and thickness of various coated layers on various textile surfaces at nanometer level. Various layers on photovoltaic fibres become clearly visible with 50000X magnification. The thickness of the layers can be seen from SEM photographs by bright interface line between the polymer anode and the photoactive layer.

7.2 Current and voltage

In order to characterize the Photovoltaic fibres open circuit voltage, short circuit current density, current and voltage at the maximum power point under an illumination of 100 mW/cm^2 are carried out.

In order to calculate the Photovoltaic efficiency of Photovoltaic textiles, current verses voltage study is essential. To achieve this target a computer controlled sourcemeter equipped with a solar simulator under a range of illumination power is required with proper calibration. All photoelectrical characterizations are advised to conduct under nitrogen or argon atmosphere inside a glove box to maintain the preciseness of observations. The overall efficiency of the PV devices can be representing by following equation.

$$\eta = \frac{V_{oc} \times I_{sc} \times FF}{P_{in}} = \frac{P_{out}}{P_{in}} \tag{1}$$

Where

V_{oc} is the open circuit voltage (for l=0) typically measured in volt (V)

I_{sc} is the short circuit current density (for V=0) in ampere / square meter (A/m^2)

P_{out} is the output electrical power of the device under illumination

P_{in} the incident solar radiation in (watt/meter^2) W/m^2

FF is the fill factor and can be explained by the following relationship:

$$FF = \frac{I_{mpp} \times V_{mpp}}{I_{sc} \times V_{oc}} \tag{2}$$

where,

V_{mpp} voltage at the maximum power point (MPP)

I_{mpp} is the current at the maximum power point (MPP)

Where the product of the voltage and current is maximized

To assure an objective measurement for precise comparison of various photovoltaic devices, characterization has to be performed under identical conditions.

An European research group has used Keithley 236 source measure unit in dark simulated AM 1.5 global solar conditions at an intensity of 100mW cm^{-2}. The solar simulator unit made by K.H. Steuernagel Lichttechnik GmbH was calibrated with the help of standard crystalline silicon diode. PV fibres were illuminated through the cathode side and I-V characteristics were measured. The semi-logarithmic I-V curves demonstrate the current density versus voltage behaviour of photovoltaic fibres under various conditions. It gives a comparative picture of voltage Vs current density as a function of various light intensities.

Durisch et al., (1996) has developed a computer based testing instrument to measure the performance of solar cells under actual outdoor conditions. This testing system consist a suntracked specimen holder, digital multimeters, devices to apply different electronic loads and a computer based laser printer. Pyranometers, pyrheliometers and a reference cell is used to measure and record the insulation. This instrument is able to test wide dimensions of photovoltaic articles ranging from 3mm X 3mm to 1 meter X 1.5meter. The major part of world's energy scientist community predicts that photovoltaic energy will play a decisive role in any sustainable energy future[63].

7.3 Electroluminescence

The institute for Solar Energy Research Hameln (ISFH) Emmerthal Germany introduced a new technique to characterize the solar cells based on electroluminescence. Electroluminescence can be defined as the emission of light resulting from a forward bias voltage application to the solar cells. The electrons recombine radioactively which are injected into the solar cells by transferring their extra energy to an emitted photon with available holes. The consequence of the electron and hole concentration is able to represent the intensity of the luminescence radiation. A powerful charge coupled device (CCD) camera is used to capture the images of intensity distribution of the luminescence radiation. Generally, actual solar cells offer inhomogeneous electroluminescence images but for an ideal solar cell it must be homogeneous. A cooled 12 CCD camera is used to capture electroluminescence images. The flexibility to adjust the distance between the camera and the solar cells offers the potential to analyse wide variety of solar cells.

7.4 Fill factor

The quality of solar cells is measured in terms of fill factor. The fill factor for a ideal solar cell is one but as internal resistance of solar cell becomes large or bad contact becomes between layers, fill factor reduces. The fill factor of textile based photovoltaics remains low due to bad quality of electrodes and/or poor contact between different layers of materials[64]. It can be improved towards unit by selection of appropriate textile substrate and further optimization process parameters and processes.

7.5 Mechanical characterization

Textile substrates are subjected to different stresses under various situations. Hence usual tensile characterization is essential for photovoltaic textiles. For tensile testing of PV fibres, the constant rate of extension (CRE) based tensile testing machines are used at 1 mm per minute deformation rate using Linear Variable Differential Transformers (LVDT) displacement sensor. Fracture phenomenon is recorded by means of high resolution video camera integrated with tensile testers.

To study about the adhesion and crack formation in coating on textile structures, generally 30 mm gauge length is used in case of photovoltaic fibres. Fibre strength measuring tensile tester, integrated with an appropriate optical microscope to record the images of specimen at an acquisition rate of about one frame per second is used to record the dynamic fracture of PV fibres. Different softwares are available to analyse the image data like PAXit, Clemex, and Digimizer etc.

7.6 Absorption spectra of solid films

Various spectrophotometers like Varian Carry 3G UV-Visible were used to observe the ultraviolet visible absorption spectra of photovoltaic films. The thin films are prepared to study the absorption spectrum of solid films. In a typical study, a thin film was prepared by spin coating of solution containing 10 mg of P3HT and 8mg of PCBM and 4.5 mg of MDMO-PPV and 18 mg of PCBM (in case of 1:4)/ml with chlorobenzene as solvent. A typical absorption spectra of MDMO-PPV:PCBM and P3HT: PCBM is illustrated in Fig 10.

7.7 X-ray diffraction of photovoltaic structures

Crystallization process is very common phenomenon that takes place during photovoltaic structure development. The content of crystalline and amorphous regions in photovoltaic structures influences the photoactivity of photovoltaic structures. X-ray diffraction technique is capable to characterize the amount of total crystallinity, crystal size and crystalline orientation in photovoltaic structures.

Presently, thin film photovoltaics are highly efficient devices being developed in different crystallographic forms: epitaxial, microcrystalline, polycrystalline, or amorphous. Critical structural and microstructural parameters of these thin film photovoltaics are directly related to the photovoltaic performance. Various X-ray techniques like x-ray diffraction for phase identification, texture analysis, high-resolution x-ray diffraction, diffuse scattering, x-ray reflectivity are used to study the fine structure of photovoltaic devices[65].

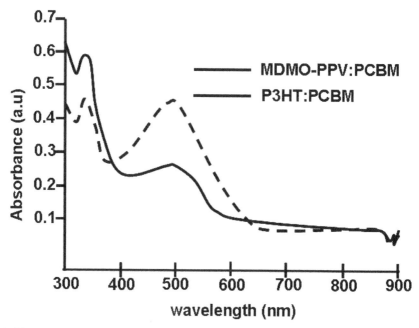

Fig. 10. Absorption spectra[57] for solutions of P3HTPCBM and MDMO-PPV-PCBM (with permission)

7.8 Raman spectroscopy

The Raman Effect takes place when light rays incidents upon a molecule and interact with the electron cloud and the bonds of that molecule. Spontaneous Raman effect is a form of scattering when a photon excites the molecule from the ground state to a virtual energy state. When the molecule relaxes it emits a photon and it returns to a different rotational or vibrational state.

Raman spectroscopy is majorly used to confirm the chemical bonds and symmetry of molecules. It provides a fingerprint to identify the molecules. The fingerprint region for organic molecules remains in the (wavenumber) range of 500–2000 cm⁻¹. Spontaneous Raman spectroscopy is used to characterize superconducting gap excitations and low frequency excitations of the solids. Raman scattering by anisotropic crystal offers information related to crystal orientation. The polarization of the Raman scattered light with respect to the crystal and the polarization of the laser light can be used to explore the degree of orientation of crystals[66].

The in situ morphological and optoelectronic changes in various photovoltaic materials can be observed by observing the changes in the Raman and photoluminescence (PL) feature with the help of a spectrometer. Various spectrums can be recorded at a definite integration time after avoiding any possibility of laser soaking of the sample[67].

8. Some facts about the photovoltaic textiles

- To achieve a highly efficient photovoltaic device, solar radiation needs to be efficiently absorbed. In case of solar cell the absorption of light causes electron hole pairs which are split into free carriers at the interface between the donor and the acceptor material.
- Active areas for photovoltaic fibres are generally found between 4 and 10mm².
- The power conversion efficiency of the MDMO-PPV:PCBM based photovoltaic fibre was higher than the P3HT:PCBM based photovoltaic fibres
- Due to circular cross-sectional shape of photovoltaic fibres, the light is absorbed at different angles
- Generally the photoactive layer thickness remain approximately between 280-350nm. A thick film can absorb more light compared to a thin film. By the increase of film thickness, the electrical field and the number of charge carriers decrease and consequently a decrease in the external quantum efficiency of the devices is observed. Although, the film thickness is restricted in presence of low-charge carrier. The optimum thickness is required to provide both maximum light absorption and maximum charge collection at the same fraction of moment. Optimization of thickness of various layers of photovoltaic fibres provides the possibility to increase the power conversion efficiency of polymer-based solar cells.
- The thickness of the layers for optimal photovoltaic fibre can be controlled by solution concentration and dipping time.
- Photovoltaic fibre based organic solar cells can be curled and crimped without losing any photovoltaic performance from their structure.
- Low power conversion efficiency of photovoltaic textiles is the real challenge in this field and can be improved by significant improvement in existing photovoltaic material and techniques. In case of organic solar cells, the optical band gap is very critical and it must be as narrow as possible because the polymers with narrow band gap are able to absorb more light at longer wavelengths, such as infrared and near-infrared. Hence low band gap polymers (<1.8 eV) can be used as better alternative for higher power harvesting efficiency in future if they are sufficiently flexible[68,69].
- The incorporation of C60 barrier layer can improve the performance of photovoltaic textiles.
- Generally the performance of freshly made photovoltaic textiles was found best because cell degradation happens fast when sun illumination takes place in absence of O_2 barriers.
- The self life of polymer based photovoltaics is short under ambient conditions[70].

9. Photovoltaic textile, developments at international level

The incorporation of polymer photovoltaics into textiles was demonstrated by Krebs et a., (2006) by two different strategies. Simple incorporation of a polyethyleneterphthalate (PET) substrate carrying the polymer photovoltaic device prepared by a doctor blade technique necessitated the use of the photovoltaic device as a structural element[71].

The total area of the device on PET was typically much smaller than the active area due to decorative design of aluminium electrode. Elaborate integration of the photovoltaic device into the textile material involved the lamination of a polyethylene (PE) film onto a suitably transparent textile material that was used as substrate. Plasma treatment of PE-surface allowed the application of a PEDOT electrode that exhibited good adherence. Screen printing of a designed pattern of poly 1,4-(2-methoxy-5-(2-ethylhexyloxy) phenylenevinylene (MEH-PPV) from chlorobenzene solution and final evaporation of an aluminum electrode completed the manufacturing of power generating device. The total area of the textile device was 1000 cm^2 (25cm x 40cm) while the active area (190 cm^2) was considerably smaller due to the decorative choice of the active material.

Konarka Inc. Lowell, Mass., U.S.A demonstrated a successful photovoltaic fiber. Presently, a German company is engaged with Ecole Polytechnique Fédérale de Lausanne (EPFL) to optimize the fiber properties and weave it into the power-generating fabric. Solar textiles would able to generate renewable power generation capabilities. The photovoltaic fibres are able to woven in fabric form rather than attached or applied on other surfaces where integration remains always susceptible. The structures woven by photovoltaic fibres are able to covert into fabric, coverings, tents and garments.

Patterned photovoltaic polymer solar cells can be incorporated on PET clothing by sewing through the polymer solar cell foil using an ordinary sewing machine. Connections between cells were made with copper wire that could also be sewn into the garment. The solar cells were incorporated into a dress and a belt as shown in Fig.11 (Tine Hertz).

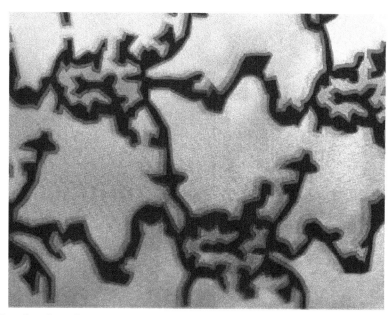

Fig. 11. Textile solar cell pattern designed by Tine Hertz and Maria Langberg of Danmarks Designkole

Shafarman et al., (2003) demonstrated thin film solar cells by using CuInGaSe$_2$ photovoltaic polymers and this film is more suitable for patching onto clothing into different patterns[72]. The polymer photovoltaics technology is in its infancy stage and many gaps need to be bridged before commercialization. Prototype printing machines are useful to apply PVs on textile surface into decorative pattern as shown in Fig. 11, 12,13.

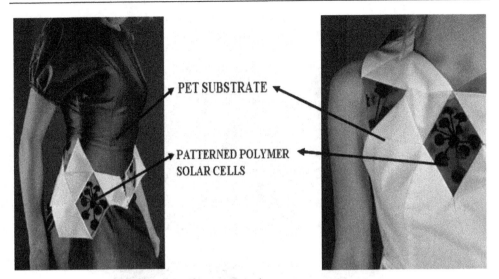

Fig. 12. Patterned polymer cell (with permission)

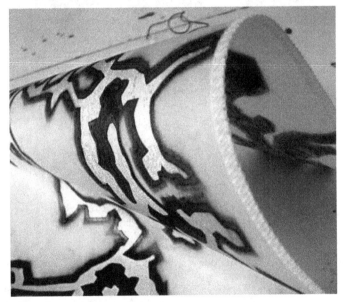

Fig. 13. Photovoltaic decorative patterns

Massachusetts Institute of Technology (MIT) Cambridge, Massachusetts revealed that the integration of solar cell technology in architecture creates designs for flexible photovoltaic materials that may change the way buildings receive and distribute energy. Sheila Kennedy of (MIT) used 3-D modeling software for her solar textiles designs, generating membrane-like surfaces that can become energy-efficient cladding for roofs or walls[73]. Solar textiles may also be used like tents as shown in Fig. 14.

Fig. 14. Photovoltaic textile as a tent (with permission)

Fig. 15. A typical example of photovoltaic textile (with permission)

- Commission for Technology and Innovation (CTI) Switzerland also exhibited a keen interest in the development of photovoltaic textiles.
- Thuringian Institute of Textiles and Plastics Research (TITK) registered their remarkable presence in order to develop photovoltaic textiles[74].
- J Wilson and R Mather have created Power Textiles Ltd, a spin-off from Heriot-Watt University, Scotland to develop a process for the direct integration of solar cells on textiles.
- Konarka is developing solar photovoltaic fabric with joint effort of the university Ecole Polytechnique Fédérale de Lausanne (EPFL), Switzerland. Konarka has claimed that

they can produce a photovoltaic fiber. Presently, the Company is working with EPFL to optimize the fiber structure and weave it into the first power-generating fabric. Solar textiles would open up additional application areas for photovoltaics since renewable power generation capabilities can be tightly integrated

- In 2002, Konarka became the first company in the United States to license Dr. Michael Grätzel's dye-sensitized solar cell technology, which augmented its own intellectual property.
- Thuringian Institute of Textiles and Plastics Research (TITK), Breitscheidstraße Rudolstadt Germany, is a technically-oriented research institute, carrying out fundamental and applied research on PV textiles suitable to easily commercialize. The institute supports small and medium-sized enterprises in their innovation works with interdisciplinary scientific knowledge, innovative ideas, and knowledge of the industry and provision of modern technical infrastructure.
- Professor John Wilson and Dr Robert Mather of School of Textiles and Design, formerly the Scottish College of Textiles have created Power Textiles Ltd, a spin-off from Heriot-Watt University, to develop a process for the direct integration of solar cells on textiles.
- In a research work at American Institute of Physics, multiwall carbon nanotubes are introduced into poly(3-hexylthiophene) and [6,6] phenyl C_{61} butyric acid methyl fullerene, bulk heterojunction organic photovoltaic devices after appropriate chemical modification for compatibility with solution processable photovoltaics. To overcome the problem of heterogeneous dispersion of carbon nanotubes in organic solvents, multiwall CNT are functionalized by acid treatment. Pristine and acid treated multiwall carbon nanotubes have been incorporated into the active layer of photovoltaic polymers which results a fill factor of 0.62 and power harvesting efficiency of 2.3% under Air Mass 1.5 Global[75].
- Dephotex is going to develop photovoltaic textiles based on novel fibre under collaboration with European Union.
- Photovoltaic tents are developed by integration of flexible solar panels made by thin film technology by patching on tent fabric surface. The solar cells can run ventilation systems, lighting and other critical electrical functions, avoiding the need for both generators and the fuel to run them.

The integration of photovoltaic technology with UV absorption technology will open very smart passages to new product development. However, the above opinion is only a hypothesis of author. The textile materials which are stable against ultraviolet rays are more suitable to work as basic substrate. However, the production and integration of photovoltaic fibres into fabric form will solve many problems concerned about simple incorporation of a polymer photovoltaic on a textile substrate directly or by lamination of a thin layer of PVs onto textile material followed by plasma treatment and application of a PEDOT electrode onto the textile materials.

10. Conclusions

The incorporation of polymeric photovoltaics into garments and textiles have been explored new inroads for potential use in "intelligent clothing" in more smart ways. Incorporation of organic solar cells into textiles has been realized encouraging performances. Stability issues need to be solved before commercialization of various photovoltaic textile manufacturing techniques. The functionality of the photovoltaic textiles does not limited by mechanical stability of photovoltaics. Polymer-based solar cell materials and manufacturing techniques

are suitable and applicable for flexible and non-transparent textiles, especially tapes and fibers, with transparent outer electrodes.
The manufactured photovoltaic fibres may also be utilized to manufacture functional yarns by spinning and then fabric by weaving and knitting. Fibres and yarns subjected to various mechanical stresses during spinning, weaving and knitting may possibly damage the coating layers of photovoltaic fibres. These sensitive and delicate structures must be protected by applying special protective layers by noble coating techniques to produce photovoltaic textiles. Photovoltaic tents, curtains, tarpaulins and roofing are available to utilize the solar power to generate electricity in more green and clean fashion.

11. References

[1] Aernouts, T. 19th European Photovoltaics Conference, June 7–11, Paris, France, 2004.
[2] Lund P D Renewable energy 34, 2009, 53
[3] Yaksel I Renewable energy 4, 2008, 802
[4] European photovoltaic Ind. Asso. Global market outlook for photovoltaic until 2012. www.epia.org
[5] Gunes S, Beugebauer H and Saricftci N S Chem Rev. 107, 2007, 1324
[6] Coakley KM, ,cGehee M D, Chem Mater. 16,2004,4533
[7] Organic Photovoltaics: Mechanisms, Materials, and Devices, ed. S.-S. Sun and N. S. Sariciftci, Taylor & Francis, London, 2005.
[8] Organic Photovoltaics: Concepts and Realization, ed. C. J. Brabec, V. Dyakonov, J. Parisi and N. S. Sariciftci, Springer-Verlag, Heidelberg, 2003.
[9] Tang, C.W. Two-layer organic photovoltaic cell. Applied Physics Letters, 48, (1986) 183.
[10] Sariciftci, N.S., Smilowitz, L., Heeger, A.J. and Wudl, F. (1992) photoinduced electron transfer from a conducting polymer to buckminsterfullerene. Science, 258, 1992, 1474.
[11] Spanggaard, H. and Krebs, F.C. (2004) A brief history of the development of organic and polymeric photovoltaics. Solar Energy Materials and Solar Cells, 83, 125.
[12] Granstrom, M., Petritsch, K., Arias, A.C., Lux, A., Andersson, M.R. and Friend, R.H. (1998) Laminated fabrication of polymeric photovoltaic diodes. Nature, 395, 257.
[13] Yu J W, Chin B D, Kim J K and Kang N S Patent IPC8 Class: AH01L3100FI, USPC Class: 136259, 2007
[14] http://www.coatema.de/ger/downloads/veroeffentlichungen/news/0701_Textile%20 Month_GB.pdf
[15] Krebs F C., Fyenbo J and Jørgensen Mikkel "Product integration of compact roll-to-roll processed polymer solar cell modules: methods and manufacture using flexographic printing, slot-die coating and rotary screen printing" J. Mater. Chem., 20, 2010, 8994-9001
[16] Krebs F C., Polymer solar cell modules prepared using roll-to-roll methods: Knife-over-edge coating, slot-die coating and screen printing Solar Energy Materials and Solar Cells 93, (4), 2009, 465-475
[17] Luo P, Zhu Cand Jiang G Preparation of CuInSe$_2$ thin films by pulsed laser deposition the Cu–In alloy precursor and vacuum selenization, Solid State Communications, 146, (1-2), 2008, 57-60
[18] Solar energy: the state of art Ed: Gordon J Pub: James and James Willium Road London, 2001
[19] Loewenstein T., Hastall A., Mingebach M., Zimmermann Y., Neudeck A. and Schlettwein D., Phys. Chem. Chem. Phys., 2008, 10,1844.
[20] Lincot D. and Peulon S., J. Electrochem. Soc., 1998, 145, 864.
[21] Bedeloglua A,*, Demirb A, Bozkurta Y, Sariciftci N S, Synthetic Metals 159 (2009) 2043–2048
[22] Pettersson LAA, Ghosh S, Inganas O. Org Electron 2002;3:143.

[23] Kim JY, Jung JH, Lee DE, Joo J. Synth Met 2002;126:311.

[24] Kim WH, Ma¨kinen AJ, Nikolov N, Shashidhar R, Kim H, Kafafi ZH. Appl Phys Lett 80, (2002), 3844.

[25] Jonsson SKM, Birgerson J, Crispin X, Greczynski G, Osikowicz W, van der Gon AWD, SalaneckWR, Fahlman M. Synth. Met 1, 2003;139:

[26] Ouyang J, Xu Q, Chu C W, Yang Y,*, Lib G, Shinar Joseph S On the mechanism of conductivity enhancement in poly(3,4-ethylenedioxythiophene) : poly(styrene sulfonate) film through solvent treatment" Polymer 45, 2004, 8443–8450

[27] Grätzel M., *Nature*, 414, 338 (2001)

[28] Könenkamp R., Boedecker,K. Lux-Steiner M. C., Poschenrieder M., Zenia F., Levy-Clement C. and Wagner S., *Appl. Phys. Lett.*, 77, 2575 (2000)

[29] Boyle D. S., Govender K. and O'Brien P., *Chem Commun.*, 1, 80 (2002)

[30] http://www.caburn.com/resources/downloads/eVapcat.pdf

[31] Hoth, C.N., Choulis, S.A., Schilinsky, P. and Brabec, C.J. "High photovoltaic performance of inkjet printed polymer: fullerene blends" Advanced Materials, 19, 2007, 3973.

[32] M Pagliaro, G Palmisano, and R Ciriminna "Flexible Solar Cells" WILEY-VCH Verlag GmbH & Co. KGaA, Weinheim, 2008, 98-119

[33] Bundgaard E and Krebs F C Low band gap polymers for organic photovoltaics Solar Energy Materials and Solar Cells 91 (11), 2007, 954-985

[34] Kroon R, Lenes M, Jan C. Paul H, Blom W. M, Boer B de "Small Bandgap Polymers for Organic Solar Cells (Polymer Material Development in the Last 5 Years)" Polymer Reviews, 48, (3), 2008 , 531 - 582

[35] Rajahn M, Rakhlin M, Schubert M B "Amorphous and heterogeneous silicone based films" MRS Proc. 664, 2001

[36] Schubert MB, Werner J H, Mater. Today 9(42), 2006

[37] Drew C, Wang X Y, Senecal K, J of Macro. Mol. Sci Pure A 39, 2002, 1085

[38] Baps B, Eder K M, Konjuncu M, Key Eng. Mater. 206-213, 2002, 937

[39] Gratzel M, Prog. Photovolt: Res. Appl. 8, 2000, 171

[40] Bayinder M, Shapira O, Sayain Hazezewski D Viens J Abouraddy A F, Jounnopoulas A D and Fink Y Nature Mat. 4, 2005, 820

[41] Verdenelli M, Parole S, Chassagneux F, Lettof J M, Vincent H and Scharff J P, J of Eur. Ceram. Soc. 23, 2003, 1207

[42] Xie C, Tong W, Acta Mater, 53, 2005, 477

[43] Muller D, Fromm E Thin Mater. Solid Films 270, 1995, 411

[44] Hu M S and Evans A G, Acta Mater, 37, 1998, 917

[45] Yang Q D, Thouless M D, Ward S M, J of Mech Phys Solids, 47, 1999, 1337

[46] Agrawal D C, Raj R, Acta Mater, 37, 1989, 1265

[47] Rochal G, Leterrier Y, Fayet P, Manson J Ae, Thin Solid Films 437,2003, 204

[48] Park S H, Choi H J,Lee S B, Lee S M L, Cho S E, Kim K H, Kim Y K, Kim M R and Lee J K "Fabrications and photovoltaic properties of dye-sensitized solar cells with electrospun poly(vinyl alcohol) nanofibers containing Ag nanoparticles" Macromolecular Research 19(2), 2011, 142-146

[49] Ramiera J., Plummera C.J.G., Leterriera Y., Mansona J.-A.E.,_, Eckertb B., and Gaudianab R. "Mechanical integrity of dye-sensitized photovoltaic fibers" Renewable Energy 33 (2008), 314–319

[50] Hamedi M, Forchheimer R and Inganas O Nat Mat. 6, 2007, 357

[51] Bayindir M, Sorin F, Abouraddy A F, Viens J, Hart S D, Joannopoulus J D, Fink Y Nature 431, 2004, 826

[52] Yadav A, Schtein M, Pipe K P J of Power Sources 175, 2008, 909

[53] Coner O, Pipe K P Shtein M Fibre based organic PV devices Appl. Phys. Letters 92, 2008, 193306

[54] Ghas A P, Gerenser L J, Jarman C M, Pornailik J E, Appl. Phys. Lett. 86, 2005, 223503

[55] Regan R O, Gratzel M, Nature 353, 1991, 737

[56] Wang P, Zakeeruddin S M, Gratzel M and Fluorine J Chem. 125, 2004, 1241

[57] Bedeloglu A C, Demir A, Bozkurt Y and Sariciftci N S "A photovoltaic fibre design for smart textiles" Text. Res. J 80(11), 2010, 1065-1074

[58] Bedeloglu A C, Koeppe R, Demir A, Bozkurt Y and Sariciftci N S "Development of energy generating photovoltaic textile structures for smart application" Fibres and Polymers 11(3), 2010, 378-383

[59] Krebs F C, Biancardo M, Winther-Jensen B, Spanggard H and Alstrup J "Strategies for incorporation of polymer photovoltaics into garment and textiles" Sol. Ener. Mat. &Sol Cells 90, 2006, 1058-1067

[60] Neef C. J. and Ferraris J. P. MEH-PPV: Improved Synthetic Procedure and Molecular Weight Control" Macromolecules, 2000, 33 (7), pp 2311–2314

[61] Winther –Jensen B and Glejbol K "Method and apparatus for the excitation of a plasma" US Patent US6628084, Published on Sept., 9, 2003

[62] Bedeloglu A, Koeppe R, Demir A, Bozkurt Y and Sariciftci N S "Development of energy generating PV textile structure for smart applications" Fibres and Polym. 11(3), 2010, 378

[63] Durisch W, Urban J and Smestad G "Characterization of solar cells and modules under actual operating conditions" WERS 1996, 359-366

[64] Kim M S, Kim B G and Kim J "Effective Variables to Control the Fill Factor of Organic Photovoltaic Cells" ACS Appl. Mater. Interfaces, , 1 (6), 2009,1264–1269

[65] Wang W, Xia G , Zheng J , Feng L and Hao R "Study of polycrystalline ZnTe(ZnTe:Cu) thin films for photovoltaic cells" Journal of Materials Science: Materials in Electronics 18(4), 2007, 427-431

[66] Khanna, R.K. "Raman-spectroscopy of oligomeric SiO species isolated in solid methane". Journal of Chemical Physics 74 (4), (1981) 2108

[67] Miller S,   Fanchini G, Lin Y Y,   Li C, Chen   C W, Sub W F and Chhowallaa M "Investigation of nanoscale morphological changes in organic photovoltaics during solvent vapor annealing" J. Mater. Chem., 2008, 18, 306–312

[68] Perzon E, Wang X, Admassie S, Inganas O, and Andersson M R "An alternative low band gap polyflourene for optoelectronic devices" Polymer 47, 2006, 4261-4268

[69] Campos L M, Tontcheva A, Gunes S, Sonmez G, Neugebauer H, Sariciftci N S and Wudl F "Extended photocurrent spectrum of a low band gap polymer in a bulk heterojunction solar cell" Chem. Mater. 17, 2005, 4031-4033

[70] Krebs F C, Carle J E, Cruys-Bagger N, Anderson M, Lilliedal M R, Hammond M A and Hvidt S, Sol. Eng Mater. Sol. Cells 86, 2005, 499

[71] Krebs F.C Spanggaard H.. Sol. Energy Mater. Sol. Cells 83 (2004) 125

[72] Shafarman, W.N., Stolt L., in: Luque A., and   Hegedus S. (Eds.), Handbook of Photovoltaic Science and Engineering, Wiley, New York, 2003.

[73] http://www.silvaco.com.cn/tech_lib_TCAD/tech_info/devicesimulation/pdf/Solar_Cell. pdf

[74] www.titk.de/en/home/home.htm

[75] Applied Physics Letters / Volume 97 / Issue 3, 2010 / NANOSCALE SCIENCE AND DESIGN

4

Organic-Inorganic Hybrid Solar Cells: State of the Art, Challenges and Perspectives

Yunfei Zhou,
Michael Eck and Michael Krüger
University of Freiburg/Freiburg Materials Research Centre
Germany

1. Introduction

Novel photovoltaic (PV) technologies are currently investigated and evaluated as approaches to contribute to a more environmental friendly energy supply in many countries. One of the driving forces are the aims to reduce the emission of green house gases and the dependency on importing fossil energy resources from political unstable countries. Additionally the wish to replace nuclear power by greener and less threatening technologies will enhance the development of regenerative energy supply in many countries especially after the recent nuclear catastrophe at Fukushima nuclear power plant in Japan in March 2011. This will include the more rapid implementation of existing mature PV technologies but also the development and improvement of novel PV approaches such as organic PV (OPV) and dye-sensitized solar cells (DSSCs) together with new efficient strategies for energy storage and distribution to make electric power, deriving from PVs, available whenever and wherever it is needed.

The so-called 1st generation of solar cells based on e.g. bulk crystalline and polycrystalline silicon is still dominating the PV market. However, so-called 2nd generation solar cells mainly consisting out of thin film solar cells based on CdTe, Copper Indium Gallium Selenide (CIGS), and amorphous silicon gained distribution of ca. 25% in market share today worldwide. It is expected that this number will increase significantly within the next years. While for the 1st and 2nd generation solar cells commercial solar panels are available with decent power conversion efficiencies (PCEs) and lifetimes, the emerging 3rd generation solar cells such as OPV and DSSCs technologies are still in the development phase. Some commercially available products have recently entered the market such as e.g. solar bags representing niche products, which are so far not suitable for competing with traditional large scale applications of solar panels of the 1st and 2nd generations. In traditional solar panels the differences between best solar cell and average solar cell efficiencies are much smaller than for the emerging solar cell technologies with the consequence that modules of 3rd generation solar cells still suffer from too low performance. In Table 1 the best cell and module efficiencies of different PV technologies are compared. It has to be mentioned that especially for the emerging new PV technologies the average efficiencies are significantly lower than the results of the best cells.

PV Technology	Best cell PCEs	Average cell PCEs	Best module PCEs	Average module PCEs
Si (bulk)	25.0% (monocryst.) (Zhao et al., 1998) 20.4% (polycryst.) (Schultz et al., 2004) 10.1% (amorphous) (Benagli et al., 2009)	---	22.9% (monocryst.) (Zhao et al., 1997) 17.55% (polycryst.) (Schott, 2010)	14-17.5% (monocryst.) 13-15% (polycryst.) 5-7% (amorphous)
CIGS (thin film)	20.3% Jackson et al., (2011)	---	15.7% (MiaSolé, 2010)	10-14%
CdTe (thin film)	16.7% (Wu X. et al., 2001)	---	10.9% (Cunningham et al., 2000)	~10%
DSSC	11.2% (Han et al., 2006)	5-9%	5,38% (Goldstein et al., 2009)	---
OPV (thin film)	8.3% (Konarka, 2010) 8.3% (Heliatek, 2010) 8.5% (Mitsubishi, 2011)	3-5%	3.86% (Solarmer, 2009)	1-3%

Table 1. Comparison of best and average PCE values of single solar cells and modules of different PV technologies.

2. Device structures and working principle

Organic-inorganic hybrid solar cells are typically thin film devices consisting out of photoactive layer(s) between two electrodes of different work functions. High work function, conductive and transparent indium tin oxide (ITO) on a flexible plastic or glass substrate is often used as anode. The photoactive light absorbing thin film consists out of a conjugated polymer as organic part and an inorganic part out of e.g. semiconducting nanocrystals (NCs). A top metal electrode (e.g. Al, LiF/Al, Ca/Al) is vacuum deposited onto the photoactive layer finally. A schematic illustration of a typical device structure is shown in Fig. 1a. Generally there are two different structure types for photoactive layers - the bilayer structure (Fig. 1b) and the bulk heterojunction structure (Fig. 1c). The latter one is usually realized by just blending the donor and acceptor materials and depositing the blend on a substrate. In contrast to bulk inorganic semiconductors, photon absorption in organic semiconductor materials does not generate directly free charge carriers, but strongly bound electron-hole pairs so-called excitons (Gledhill et al., 2005). Since the exciton diffusion lengths in conjugated polymers are typically around 10-20 nm (Halls et al., 1996) the optimum distance of the exciton to the donor/acceptor (D/A) interface, where charge transfer can take place and excitons dissociate into free charge carriers, should be in the same length range. Therefore the bulk-heterojunction structure was introduced where the electron donor and acceptor materials are blended intimately together (Halls et al., 1995). The interfacial area is dramatically increased and the distance that excitons have to travel to reach the interface is reduced. After exciton dissociation into free charge carriers, holes and electrons are transported via polymer and NC percolation pathways towards the respective electrodes. Ideally, an interdigital donor acceptor configuration would be a perfect structure for efficient exciton dissociation and charge transport (Fig. 1d). In such a structure, the distance from exciton generation sites, either in the donor or the acceptor phase, to the D/A interface would be in the range of the exciton diffusion length. After exciton dissociation, both holes and electrons will be transported within their pre-structured donor or acceptor phases along a direct percolation pathway to the respective electrodes. This interdigital structure can be realized by various nanostructuring approaches, which will be discussed in detail later in the section 6.2.2.

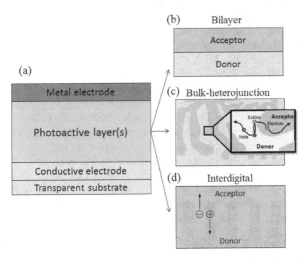

Fig. 1. Schematic illustration of typical device structures for hybrid solar cells.

In hybrid solar cells, photocurrent generation is a multistep process. Briefly, when a photon is absorbed by the absorbing material, electrons are exited from the valance band (VB) to the conduction band (CB) to form excitons. The excitons diffuse to the donor/acceptor interface where charge transfer can occur leading to the dissociation of the excitons into free electrons and holes. Driven by the internal electric field, these carriers are transported through the respective donor or acceptor material domains and are finally collected at the respective electrodes. To sum up, there are four main steps: photon absorption, exciton diffusion, charge separation as well as charge carrier transport and collection. The physics of organic/hybrid solar cells is reviewed in detail elsewhere (Greenham, 2008; Saunders & Turner, 2008).

3. Donor-acceptor materials

Due to the decreased size of NCs down to the nanometer scale, quantum effects occur, thus a number of physical (e.g. mechanical, electrical, optical, etc.) properties change when compared to those of bulk materials. For example, the quantum confinement effect (Brus, 1984) can be observed once the diameter of the material is in the same magnitude as the wavelength of the electron wave function. Along with the decreasing size of NCs, the energy levels of NCs turn from continuous states to discrete ones, resulting in a widening of the band gap apparent as a blue shift in the absorption and photoluminescence (PL) spectra. In general, there are two distinct routes to produce NCs: by physical approaches where they can be fabricated by lithographic methods, ion implantation, and molecular beam deposition; or by chemical approaches where they are synthesized by colloidal chemistry in solution. Colloidal synthetic methods are widely used and are promising for large batch production and commercial applications. The unique optical and electrical properties of colloidal semiconductor NCs have attracted numerous interests and have been explored in various applications like light-emitting diodes (LEDs) (Kietzke, 2007), fluorescent biological labeling (Bruchez et al., 1998), lasers (Kazes et al., 2002), and solar cells (Huynh et al., 2002). Colloidal NCs synthesized in organic media are usually soluble in common organic solvents thus they can be mixed together with conjugated polymers which are soluble in the same solvents. With suitable band gap and energy levels, NCs can be incorporated into conjugated polymer blends to form so-called bulk-heterojunction hybrid solar cells (Borchert, 2010; Reiss et al., 2011; Xu & Qiao, 2011; Zhou, Eck et al., 2010). CdS, CdSe, CdTe, ZnO, SnO$_2$, TiO$_2$, Si, PbS, and PbSe NCs have been used so far as electron acceptors. In Table 2 different donor-acceptor combinations in 3rd generation solar cells are shown together with the respective highest achieved PCEs from laboratory devices.

Bulk-heterojunction hybrid solar cells are still lagging behind the fullerene derivative-based OPVs in respect of device performance. Nevertheless, they have the potential to achieve better performance while still maintaining the benefits such as potentially low-cost, thin and flexible, and easy to produce. By tuning the diameter of the NCs, their band gap as well as their energy levels can be varied due to the quantum size effect. Furthermore, quantum confinement leads to an enhancement of the absorption coefficient compared to that of the bulk materials (Alivisatos, 1996). As a result, in the NCs/polymer system, both components have the ability to absorb incident light, unlike the typical polymer/fullerene system where the fullerene contributes very little to the photocurrent generation (Diener & Alford, 1998; Kazaoui & Minami, 1997). In addition, NCs can provide stable elongated structures on the length scale of 2-100 nm with desirable exciton dissociation and charge transport properties (Huynh et al., 2002).

Donor	Acceptor	PCE(%)	Reference
Polymer	C$_{60}$ derivative	8.3	(Konarka, 2010)
Polymer	CdSe Tetrapods	3.19	(Dayal et al., 2010)
Polymer	Polymer	2.0	(Frechet et al., 2009)
Small molecule	Small molecule	8.3	(Heliatek, 2010)
Dye	TiO$_2$	11.2	(Han et al., 2006)

Table 2. Donor-acceptor combinations and best PCEs of 3rd generation solar cells.

Fig. 2 illustrates commonly used donor and acceptor materials in bulk-heterojunction hybrid solar cells. The conjugated polymers usually act as electron donors and semiconductor NCs with different shapes such as spherical quantum dots (QDs), nanorods (NRs) and tetrapods (TPs) as well as the C$_{60}$ derivative PCBM as electron acceptor materials.

In Fig. 3 the energy levels (in eV) of commonly used conjugated polymers as donors and NCs as acceptors for bulk-heterojunction hybrid solar cells are summarized and compared. The Fermi levels of the electrodes and the energy levels of PCBM are shown as well. The variation of the values for the energy levels are deriving from different references and are due to different applied measurement methods for extracting the respective values of the lowest unoccupied molecular orbitals and highest occupied molecular orbitals (HOMO-LUMO) levels such as cyclic voltammetry (CV), X-ray photoelectron spectroscopy (XPS), ultra-violet photoelectron spectroscopy (UPS). The data for the respective HOMO-LUMO levels have been extracted from various references which are given in a recent review article (Zhou, Eck et al., 2010).

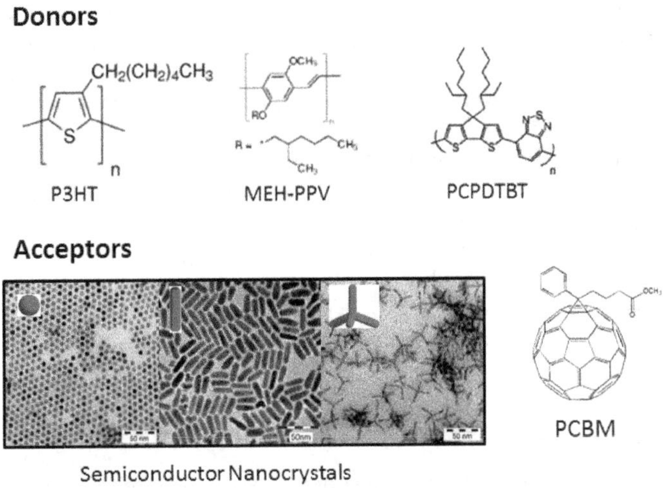

Donors

P3HT   MEH-PPV   PCPDTBT

Acceptors

PCBM

Semiconductor Nanocrystals

Fig. 2. Up: Chemical structures of commonly used conjugated polymers as electron donors for bulk-heterojunction hybrid solar cells. Shown are Poly(3-hexylthiophene-2,5-diyl) (P3HT), Poly[2-methoxy-5-(2-ethylhexyloxy)-1,4-phenylenevinylene](MEH-PPV), and Poly[2,6-(4,4-bis-(2-ethylhexy)-4H-cyclopenta[2,1-b;3,4-b]-dithiophene)-alt-4,7-(2,1,3-benzothiadiazole)](PCPDTBT). Down: Differently shaped semiconductor NCs as well as the chemical structure of [6,6]-Phenyl C_{61} butyric acid methyl ester (PCBM) as electron acceptors.

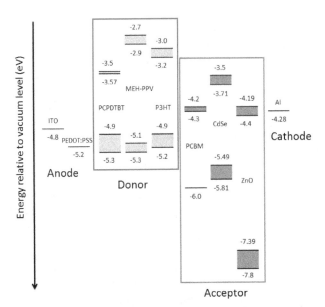

Fig. 3. Energy levels (in eV) of commonly used conjugated polymers as electron donors and NCs as electron acceptors in bulk-heterojunction hybrid solar cells.

Energy levels of donor and acceptor materials should match for efficient charge separation at the D/A interface. PL spectroscopy is a simple and useful method to investigate if a material combination can be an appropriate D/A system (Greenham et al., 1996). Because pure polymers such as P3HT and MEH-PPV exhibit a strong PL behaviour, its PL intensity is quenched by the addition of NCs with matching energy levels. This is an indication that charge transfer occurs from polymer to NCs. However, the observation of PL quenching is not necessarily a proof of charge separation within the D/A system because Förster resonance energy transfer (FRET) could also happen from larger band gap materials to smaller band gap materials, leading to strong PL quenching as well (Greenham et al., 1996). Therefore, additional methods such as photoinduced absorption (PIA) spectroscopy and light-induced electron spin resonance (L-ESR) spectroscopy are used in order to exclude PL quenching due to FRET. A detailed review on these two methods has been recently published (Borchert, 2010).

4. CdSe NCs based hybrid solar cells

CdSe NCs were the first NCs being incorporated into solar cells which still exhibit the highest PCEs compared to devices with NCs from other materials, and are still under extensive studies for utilization in hybrid solar cells. CdSe NCs have some advantages: they absorb at a useful spectral range for harvesting solar emission from 300 nm to 650 nm, they are good electron acceptors in combination with conjugated polymers, and the synthetic methods for their synthesis are well-established. The incorporation of CdSe spherical quantum dots into polymer for hybrid solar cells was firstly reported in 1996 (Greenham et al., 1996). At a high concentration of NCs of around 90% by weight (wt%), external quantum efficiencies (EQE) up to 10% were achieved, indicating an efficient exciton dissociation at the polymer/NCs interface. Although the phase separation, between the polymer and the NCs was observed to be in the range of 10-200 nm, the PCEs of devices were very low of about 0.1%. This was attributed to an inefficient electron transport between the individual NCs. After different shapes of NCs were synthetically available (Peng X. G. et al., 2000), different elongated CdSe structures were utilized in hybrid solar cells as electron acceptor materials.

Meanwhile numerous approaches were published regarding the synthesis of various morphologies and structures of CdSe NCs such as QDs, NRs and TPs and their application in hybrid solar cells. A significant advance was reported in 2002 (Huynh et al., 2002), when efficient hybrid solar cells based on elongated CdSe NRs and P3HT were obtained. Elongated NRs were used for providing elongated pathways for effective electron transport. Additionally, P3HT was used as donor material instead of MEH-PPV since it has a comparatively high hole mobility and absorbs at a longer wavelength range compared to PPV derivatives (Schilinsky et al., 2002). By increasing the NRs length, improved electron transport properties were demonstrated resulting in an improvement of the EQE. The optimized devices consisting out of 90wt% pyridine treated nanorods (7 nm in diameter and 60 nm in length) and P3HT exhibited an EQE over 54% and a PCE of 1.7%. Later on, 1,2,4-trichlorobenzene (TCB), which has a high boiling point, was used as solvent for P3HT instead of chlorobenzene. It was found that P3HT forms fibrilar morphology when TCB was used as solvent providing extended pathways for hole transport, which resulted in improved device efficiencies up to 2.6% (Sun & Greenham, 2006). Further improvement was achieved by using CdSe TPs, since TPs always have an extension perpendicular to the electrode for more efficient electron transport in comparison to NRs which are preferentially oriented more parallel to the electrode (Hindson et al., 2011). Devices based on pyridine treated CdSe TPs exhibited efficiencies up to 2.8% (Sun et al., 2005). Recently, by using the lower band gap polymer PCPDTBT, which can absorb a higher fraction of the solar emission, an efficiency of 3.19% was reported (Dayal et al., 2010). This value is up to date the highest efficiency for colloidal NCs based bulk-heterojunction hybrid solar cells.

Elongated or branched NCs in principal can provide more extended and directed electrical conductive pathways, thus reducing the number of inter-particle hopping events for extracting electrons towards the electrode. However, device performance does not only benefit from the shape of the NCs, but also from their solubility and surface modification which influence significantly the charge transfer and carrier transport behavior. Despite the relatively high intrinsic conductivity within the individual NCs, the electron mobility through the NC network in hybrid solar cells is quite low, which could be mainly attributed to the electrical insulating organic ligands on the NC surface. Ginger *et al.* have investigated charge injection and charge transfer in thin films of spherical CdSe NCs covered with TOPO ligand sandwiched between two metal electrodes (Ginger & Greenham, 2000). Very low electron mobilities in the order of 10^{-5} $cm^2V^{-1}s^{-1}$ were measured, whereas the electron mobility of bulk CdSe is in the order of 10^2 $cm^2V^{-1}s^{-1}$ (Rode, 1970). In most cases, the ligands used for preventing aggregation during the growth of the NCs contain long alkyl chains, such as oleic acid (OA), trioctylphosphine oxide (TOPO) or hexadecylamine (HDA), form electrically insulating layers preventing an efficient charge transfer between NCs and polymer, as well as electron transport between the individual NCs (Greenham et al., 1996; Huynh et al., 2003). In order to overcome this problem, post-synthetic treatment on the NCs has been investigated extensively. Fig. 4 shows two general strategies of post-synthetic treatment on NCs for improving the performance of hybrid solar cells – ligand exchange from original long alkyl ligands to shorter molecules e.g. pyridine, and chemical surface treatment and washing for reducing the ligand shell. A combination of ligand shell reduction and ligand exchange afterwards might further improve the solar cell performance by enhancing the electron transport in the interconnected NC network.

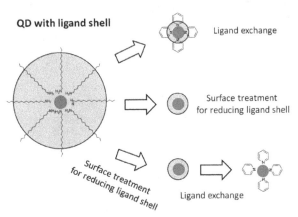

Fig. 4. Schematic illustration of two post-synthetic QD treatment strategies to enhance the PCEs in hybrid solar cells: ligand exchange (up) and reduction of the ligand surface of QDs by applying a washing procedure (middle). A combination of the two approaches might be beneficial for further enhancing the performance of hybrid solar cells (down).

Pyridine ligand exchange is the most commonly used and effective postsynthetic procedure so far, leading to the state-of-the-art efficiencies for hybrid solar cells (Huynh et al., 2002). Generally, as-synthesized NCs are washed by methanol several times and consequently refluxed in pure pyridine at the boiling point of pyridine under inert atmosphere overnight. This pyridine treatment is believed to replace the synthetic insulating ligand with shorter and more conductive pyridine molecules.

Treatments with other materials such as chloride (Owen et al., 2008), amine (Olson et al., 2009), and thiols (Aldakov et al., 2006; Sih & Wolf, 2007) were also investigated. Aldakov et al. systematically investigated CdSe NCs modified by various small ligand molecules with nuclear magnetic resonance (NMR), optical spectroscopy and electrochemistry, although their hybrid devices exhibited low efficiencies (Aldakov et al., 2006). Olson *et al.* reported on CdSe/P3HT blended devices exhibiting PCEs up to 1.77% when butylamine was used as a shorter capping ligand for the NCs (Olson et al., 2009). In an alternative approach, shortening of the insulating ligands by thermal decomposition was demonstrated and led to a relative improvement of the PCEs of the CdSe/P3HT-based solar cells (Seo et al., 2009).

However, NCs after ligand exchange with small molecules tend to aggregate and precipitate out of the organic solvent because long alky chain ligands are replaced (Huynh et al., 2002; Huynh et al., 2003), resulting in difficulties to obtain stable mixtures of NCs and polymer. Recently, a new strategy for post-synthetic treatment on spherical CdSe QDs was demonstrated (Zhou, Riehle et al., 2010), where the NCs were treated by a simple and fast hexanoic acid-assisted washing procedure. One advantage of avoiding the exchange of the synthesis capping ligands is that the QDs retain a good solubility after acid treatment, resulting in reproducible performance as well as allowing a high loading of the CdSe QDs in the blend, which is preferable for an efficient percolation network formation during the annealing step of the photoactive composite film. Devices with optimized ratios of QDs to P3HT exhibited reproducible PCEs up to 2.1% after spectral mismatch correction (Zhou, Eck et al., 2010) (Fig. 5a). This is the highest reported value for a CdSe QD / P3HT based hybrid solar cell so far. It is notable that the FF is relatively high up to 0.54, implying a good charge carrier transport capability in the devices. A simple reduced ligand sphere model was proposed to explain the possible reason for improved photovoltaic device efficiencies after acid treatment as shown in Fig. 5b (Zhou, Riehle et al., 2010). By the assistance of hexanoic acid this "immobilized" insulating spheres formed by HDA ligands are effectively reduced in size due to the salt formation of HDA. This organic salt is also much more easily dissolved in the supernatant solution than unprotonated HDA and can be separated easily from the QDs by subsequent centrifugation.

In addition, extended investigations on TOP/OA capped CdSe QDs suggested that the hexanoic acid treatment is also for this ligand system applicable for improving the device performance. Although these two kinds of QDs have different sizes (5.5 nm for HDA-capped QDs and 4.7 nm for TOP/OA capped QDs) which could result in different energy levels of QDs as well, after acid treatment both devices exhibit PCEs of 2.1% (Zhou et al., 2011) as shown in Fig. 6. Furthermore, using low band gap polymer PCPDTBT, optimized devices based on acid treated TOP/OA CdSe QDs were achieved and exhibited the highest efficiency of 2.7% for CdSe QD based devices so far (Zhou et al., 2011).

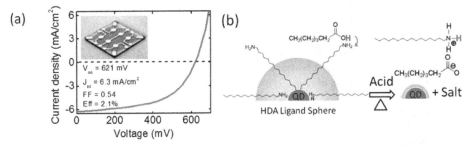

Fig. 5. (a) J-V characteristic of a hybrid solar cell device containing 87 wt% CdSe QDs and P3HT as photoactive layer under AM1.5G illumination, exhibiting a PCE of 2.1% after spectral mismatch correction (Inset: Photograph of the hybrid solar cell device structure) [Zhou, Eck et al., 2010] – Reproduced by permission of The Royal Society of Chemistry. (b) Schematic illustration of the proposed QD sphere model: an outer insulating HDA ligand sphere is supposed to be responsible for the insulating organic layer in untreated QDs directly taken out of the synthesis matrix and is effectively reduced in size by methanol washing and additional acid treatment. Reprinted with permission from [Zhou, Riehle et al., 2010]. Copyright [2010], American Institute of Physics..

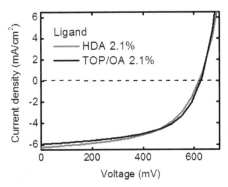

Fig. 6. Comparison of J-V characteristics of the best devices fabricated based on HDA or TOP/OA ligand capped CdSe QDs and P3HT, exhibiting similar PCEs of 2.1%.

5. Hybrid solar cells based on other NCs

Other semiconductor NCs than CdSe were also used for hybrid solar cells. ZnO NCs have attracted a lot of attention because they are less toxic than other II-VI semiconductors and are relatively easy to synthesize in large quantities. Devices based on blends of MDMO-PPV and ZnO NCs at an optimized NC content (67 wt%) presented a PCE of 1.4% (Beek et al., 2004). By using P3HT as donor polymer which has a higher hole mobility together with an in-situ synthesis approach of ZnO directly in the polymer matrix, the efficiency was optimized up to 2% using a composite film containing 50 wt% ZnO NCs (Oosterhout et al., 2009). However, because of the relatively large band gap, the contribution to the absorption of light from ZnO NCs is very low. Another disadvantage is the low solubility of ZnO NCs in solvents which are commonly used for dissolving conjugated polymers (Beek et al., 2006).

This problem of processing ZnO NCs together with polymers to obtain well-defined morphologies limits up to now the further improvement of the solar cell performance of ZnO based hybrid solar cells.

Low band gap NCs such as CdTe, PbS, PbSe, CuInS$_2$ and CuInSe$_2$ NCs are promising acceptor materials due to their ability of absorbing light at longer wavelengths which may allow an additional fraction of the incident solar spectrum to be absorbed. For instance, CdTe NCs have a smaller band gap compared to CdSe NCs, while their synthesis routes are similar to CdSe NCs (Peng & Peng, 2001). However, suitable CdTe/polymer systems have not yet been found, and reported PCEs based on CdTe/MEH-PPV are quite below 0.1% (Kumar & Nann, 2004). A systematic investigation on hybrid solar cells based on MEH-PPV blended with CdSe$_x$Te$_{1-x}$ tetropods demonstrated a steady PCE decrease from 1.1% starting from CdSe to 0.003% with CdTe (Zhou et al., 2006). The reason of the dramatically decrease in efficiency could be attributed to the possibility that energy transfer rather than charge transfer could occur from the polymer to CdTe NCs in CdTe/Polymer blends, resulting in an insufficient generation of free charge carriers (van Beek et al., 2006; Zhou et al., 2006). However there is one work reporting over 1% efficiency using vertically aligned CdTe nanorods combined with poly(3-octylthiophene) (P3OT), indicating that CdTe NCs may be useful for hybrid solar cells when the energy levels are matching to the polymers (Kang et al., 2005). Further lowering of the NC band gap could be achieved by using semiconductors such as PbS or PbSe. Watt et al. have developed a novel surfactant-free synthetic route where PbS NCs were synthesized in situ within a MEH-PPV film (Watt et al., 2004; Watt et al., 2005). CuInS$_2$ and CuInSe$_2$ which have been successfully used in inorganic thin film solar cells are promising for hybrid solar cells as well. Although an early study performed by Arici et al. (Arici et al., 2003) showed very low efficiencies <0.1%, recent progress on colloidal synthesis methods for high quality CuInS$_2$ (Panthani et al., 2008; Yue et al., 2010) might stimulate the development to more efficient photovoltaic devices. In general, using low band gap NCs as electron acceptors in polymer/NCs systems has been not successful yet, because energy transfer from polymer to low band gap NCs is the most likely outcome, resulting in inefficient exciton dissociation.

Recently it has been demonstrated that Si NCs are a promising acceptor material for hybrid solar cells due to the abundance of Si compounds, non-toxicity, and strong UV absorption. Hybrid solar cells based on blends of Si NCs and P3HT with a PCE above 1% have been reported (Liu et al., 2009). Si NCs were synthesized by radio frequency plasma via dissociation of silane, and the size can be tuned between 2 nm and 20 nm by changing chamber pressure, precursor flow rate, and radio frequency power. Devices made out of 50 wt% Si NCs, 3-5 nm in size, exhibited a PCE of 1.47% under AM1.5 G illumination which is a promising result (Liu et al., 2010).

The distribution of ligand-free NCs into the conjugated polymer matrix should be of great advantage for the resulting hybrid solar cells. This can be realized by an "in situ" synthesis approach of NCs directly in the polymer matrix. First attempts have been performed with a one pot synthesis of PbS in MEH-PPV by Watt et al. (Watt et al. 2005). Although the size distribution and concentration of synthesized NCs was not optimized, a PCE of 1.1 % was reached using this method. Liao et al. demonstrated successfully a direct synthesis of CdS nanorods in P3HT, leading to hybrid solar cells with PCEs up to 2.9% (Liao et al., 2009).

Table 3 summarized the selected performance parameters of hybrid solar cells based on colloidal NCs and conjugated polymers.

NC	Shape	Polymer	PCE(%)	Reference
CdSe	TP	PCPDTBT	3.19	(Dayal et al., 2010)
CdSe	TP	OC$_1$C$_{10}$-PPV	2.8	(Sun et al., 2005)
CdSe	QD	PCPDTBT	2.7	(Zhou et al., 2011)
CdSe	NR	P3HT	2.65	(Wu & Zhang, 2010)
CdSe	NR	P3HT	2.6	(Sun & Greenham, 2006)
CdSe	TP	APFO-3	2.4	(Wang et al., 2006)
CdSe	Hyperbranched	P3HT	2.2	(Gur et al., 2007)
CdSe	QD	P3HT	2.0	(Zhou, Riehle et al., 2010)
CdSe	QD	P3HT	1.8	(Olson et al., 2009)
CdSe	NR	P3HT	1.7	(Huynh et al., 2002)
ZnO	-	P3HT	2.0	(Oosterhout et al., 2009)
ZnO	-	P3HT	1.4	(Beek et al., 2004)
CdS	NR	P3HT	2.9	(Liao et al., 2009)
CdTe	NR	MEH-PPV	0.05	(Kumar & Nann, 2004)
CdTe	NR	P3OT	1.06	(Kang et al., 2005)
PbS	QD	MEH-PPV	0.7	(Gunes et al., 2007)
PbSe	QD	P3HT	0.14	(Cui et al., 2006)
Si	QD	P3HT	1.47	(Liu et al., 2010)

Table 3. Selected performance parameters of hybrid solar cells reported in literature based on colloidal NCs and conjugated polymers.

6. Challenges and perspectives

6.1 Extension of the photon absorption and band gap engineering

Absorption of a large fraction of the incident photons is required for harvesting the maximum possible amount of the solar energy. Generally, incident photons are mainly absorbed by the donor polymer materials and partially also from the inorganic NCs. For example in blends containing 90 wt% CdSe nanoparticles in P3HT, about 60% of the total absorbed light energy can be attributed to P3HT due to its strong absorption coefficient (Dayal et al., 2010). Using P3HT as donor polymer, hybrid solar cells with spherical QDs, NRs, and hyperbranched CdSe NCs exhibited the best efficiencies of 2.0%(Zhou, Riehle et al., 2010), 2.6%(Sun & Greenham, 2006; Wu & Zhang, 2010), and 2.2%(Gur et al., 2007), respectively. However, due to the insufficient overlap between the P3HT absorption spectrum and the solar emission spectrum (Scharber et al., 2006), further improving of the PCE values seems to be difficult to obtain with this polymer system.

Assuming that all photons up to the band gap edge are absorbed and converted into electrons without any losses (i.e. external quantum efficiency (EQE) is constant 1), crystalline silicon with a band gap of 1.1 eV can absorb up to 64% of the photons under AM1.5 G illumination, with a theoretical achievable current density J_{sc} of about 45 mA/cm^2. While in the case of P3HT having a band gap of 1.85 eV, only 27% photons can be absorbed, resulting in a maximal J_{sc} of 19 mA/cm^2. By using a low band gap polymer with a band gap of e.g. about 1.4 eV, 48% photons can be absorbed leading to a maximum J_{sc} up to 32 mA/cm^2 (Zhou, Eck et al., 2010). Nevertheless, lowering the band gap of photo-absorbing materials below a certain limit will lead to a decrease in device efficiency, because the energy of absorbed photons with a larger energy than the band gap will be wasted as the electrons and holes relax to the band edges.

Most low band gap polymers are from the material classes of thiophene, fluorene, carbazole, and cylopentadithiophene based polymers, which are reviewed in detail in several articles (Kamat, 2008; Riede et al., 2008; Scharber et al., 2006). Among those low band gap polymers, PCPDTBT (chemical structure shown in Fig.2) with a band gap of ~1.4 eV and a relatively high hole mobility up to 1.5×10^{-2} cm²V⁻¹s⁻¹ (Morana et al., 2008) appears to be an excellent candidate as a photon-absorbing and electron donating material (Soci et al., 2007). OPVs based on PCPDTBT:PC₇₀BM system achieved already efficiencies up to 5.5% (Peet et al., 2007) and 6.1%(Park et al., 2009). Recently, a bulk-heterojunction hybrid solar cell based on CdSe tetrapods and PCPDTBT was reported by Dayal et al.(Dayal et al., 2010) with an efficiency of 3.13%. Devices based on PCPDTBT and CdSe TPs, exhibited an EQE of >30% in a broad range from 350 nm to 800 nm, which is the absorption band of the polymer. It is notable that the devices reached very high J$_{sc}$ values above 10 mA/cm², indicating that the broad absorption ability of the photoactive hybrid film consequently contributes to the photocurrent. Zhou et al. reported on a direct comparison study of using PCPDTBT and P3HT as donor polymer for CdSe QDs based hybrid solar cells (Zhou et al., 2011). Fig. 7a shows the comparison of the best cells fabricated from blends of P3HT:CdSe and PCPDTBT:CdSe. The PCPDTBT based device showed a considerable enhancement of PCE to 2.7% compared to the P3HT based device mainly due to the increase of J$_{sc}$. Fig. 7b shows the EQE spectrum of photovoltaic devices comparing the two different polymers. The PCPDTBT based device showed a broader EQE spectrum from 300 nm to 850 nm, and considerable photocurrent contribution from the QDs was observed at 400 nm region where the QD absorption is strong. This implies that both components of the PCPDTBT:CdSe system contribute to the absorption of incident photons and to the photocurrent generation. The energy levels of donor and acceptor materials also play an important role determining the V$_{oc}$ and consequently device efficiency. The optimum LUMO offset between donor and acceptor has been investigated by many research groups. An offset energy of 0.3 eV was found to be sufficient for charge transfer (Brabec et al., 2002; Bredas et al., 2004). Therefore, fitting of the donor and acceptor energy levels as well as band gap engineering are desirable for eliminating energy losses during the charge transfer process. Scharber et al. demonstrated a relationship between PCEs of solar cells, band gaps, and the offsets between donor and acceptor LUMO levels of the donor materials.

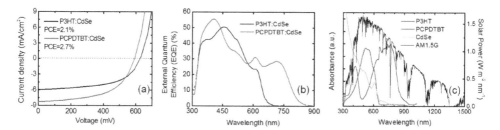

Fig. 7. (a) J-V characteristics of the best solar cells fabricated from blends of P3HT:CdSe and PCPDTBT:CdSe with a PCEs of 2.1% and 2.7% respectively. (b) EQE spectra of the P3HT:CdSe and PCPDTBT:CdSe devices (c) Absorption spectra of CdSe QDs, P3HT, PCPDTBT in thin films, in comparison with the AM1.5G solar emission spectrum.

As a result, for PCEs of devices exceeding 10%, a donor band gap <1.74 eV and a LUMO level <-3.92 eV are required (Scharber et al., 2006). In addition, Dennler et al. demonstrated that for a minimum energy offset of 0.3 eV between the donor and acceptor LUMO levels, PCEs of >10% are practical available for a donor polymer with an ideal optical band gap of ~1.4 eV (Riede et al., 2008). Recently, Xu et al. predicted the highest achievable cell efficiencies in polymer/NCs hybrid solar cells by considering the polymer band gaps and polymer LUMO energy levels (Xu & Qiao, 2011). Fig. 9 illustrates the 3D contour plots of polymer LUMO levels, polymer band gaps, and calculated device efficiencies for three representative inorganic NCs with CBs at ~4.2 eV (TiO$_2$), ~4.4 eV (ZnO) and ~3.7 eV (CdSe). Assuming all of the photons are absorbed by the polymers and the V$_{oc}$ equals to the energy offset between the polymer HOMO and the NC LUMO, device efficiencies beyond 10% can be achieved by using polymers with optimal band gaps and LUMO levels.

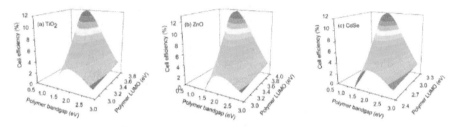

Fig. 8. 3D contour plots of polymer LUMO energy levels, polymer band gaps and cell efficiencies in a single junction solar cell structure with three representative inorganic semiconductor acceptors of (a) TiO2; (b) ZnO; and (c) CdSe. The conversion efficiencies of solar cells were calculated by assuming IPCE =65%, FF=60% under AM 1.5 with an incident light intensity of 100 mW cm^2. [Xu & Qiao, 2011] – Reproduced by permission of The Royal Society of Chemistry.

Another approach to increase the photon absorption in the active layer is to use light trapping structures as substrates or as electrodes. Light trapping can be used to overcome the problem of insufficient absorption in thin film solar cells in general (Rim et al., 2007). Nano- and microstructures of the photoactive film material can be utilized to enlarge the total pathway of incident light through the active layer. An early attempt for realizing a light trapping structure in organic solar cells was made by Roman et al. (Roman et al., 2000), who used a sub micrometer patterned grating created by lithography to mold the active layer in a way that it exhibits a cross sectional saw tooth characteristic on its surface. The subsequently deposited aluminum top electrode is then acting as a reflactive layer. Furthermore Niggemann et al. created buried nanoelectrodes (Niggemann et al., 2004) requiring an inverted cell design by using a structured aluminium coated electrode, and microprisms (Niggemann et al., 2008) as light trapping substrate for organic solar cells. So far light trapping was only applied on pure organic solar cells, but because of the similar device structure these attempts can also be applied to hybrid solar cells.

6.2 Enhancing of the charge carrier transport in hybrid solar cells

The film morphology plays a decisive role in the performance of a hybrid solar cell. Both the nano-phase separation and the charge extraction must be optimized for a highly efficient solar cell. For an optimal nano-phase separation the acceptor material must be homogeneously distributed in the blend. For optimizing the charge extraction towards the electrodes, continuous percolation pathways should exist for the charges to move towards the respective electrodes. For optimized hybrid films with suitable nano-phase separation the solvents used for the donor-acceptor blends must suit well for both the NCs and the polymer. In order to improve the dispersibility of the NCs inside the NC/polymer mixture, pyridine is added in a certain optimal concentration to the P3HT solvent chloroform. The experiment resulted in a reduction of the surface roughness of the hybrid film which was measured by AFM and in a higher EQE for hybrid devices (Huynh et al., 2003).

The crystallinity of the conjugated polymer is another important factor to consider for improving the hole extraction towards the anode of the solar cell. Therefore P3HT is a suitable material (Sharma et al., 2010). Greenham's group reported that using TCB as a solvent with a slow evaporation rate, in contrast to chloroform, is enhancing the self organization of the polymer and thereby the efficiency of a P3HT/CdSe (NR) solar cell (Sun & Greenham, 2006). Additionally during the the thermal treatment interfacial and access ligands (e.g. pyridine) are removed (Huynh et al., 2003). By treating the blend at temperatures of ca. 110°C it is reported that oxygen is removed from the P3HT (Olson et al., 2009). Erb et. al reported that the crystallinity of P3HT is improving significantly after thermal annealing which can be observed by the extension of the absorption spectra to longer wavelengths after thermal treatment of the polymer (Erb et al., 2005).

6.2.1 Visualization of the nanomorphology of thin hybrid films

An AFM analysis of the active layer of the hybrid blend reveals information about the surface topography. Here the roughness is mostly regarded as indicator for the quality of the nanophase separation of NC and polymer phases. An AFM image of the surface of a CdSe/P3HT hybrid film is shown in Fig. 9a. In addition TEM can be used for the investigation of thin hybrid films. The two dimensional image delivers information about the distribution of donor and acceptor materials in the film (Fig. 9b). Hereby the quality of the mixing and the tendency of NC aggregation as well as nanophase separation can be observed. A relatively new approach for the analysis of the nanomorphology in hybrid solar cells is the use of 3D TEM tomography, where a series of TEM images are taken of the sample subsequently at different tilt angles.

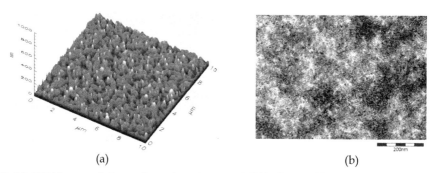

(a)   (b)

Fig. 9. (a) AFM image of the surface of a spin coated CdSe/P3HT blend film, (b) TEM of a CdSe/P3HT thin film. The white areas represent the polymer phase and the dark areas the NC phase.

With the help of a computer software a three dimensional tomographic view of the donor-acceptor blend can be achieved (Fig. 10). This method is especially well suited for hybrid solar cells because they exhibit a high contrast between the inorganic NCs and the organic polymer. The obtained visualization of the internal material distribution gives an important feedback for solar cell development.

In Fig. 10a a 3D visualization of a P3HT/ZnO thin hybrid film is shown (Oosterhout et al., 2009). The volume fraction of NCs present in the active area could be successfully extracted. Furthermore the fraction of NCs connected to the top electrode could be calculated, which was decreasing from 93% for a 57 nm thin film to 80% for a 167 nm thin film. This decrease could be correlated with the decrease of the IQE. Surprisingly despite of the better IQE, thinner films are showing a considerably coarser nanophase separation with only 60% of the fraction of the P3HT lying at a distance of 10 nm or less to the next acceptor, while this value was nearly 100% for the 167 nm thick film.

Fig.10b shows analyzed blends of OC_1C_{10}-PPV/CdSe by 3D TEM tomography (Hindson et al., 2011). It was demonstrated that the better performance of CdSe NR based devices is due to the higher connectivity between the NCs leading to a total fraction of ca. 90% NRs being connected to the top electrode while for QD based cells the fraction of connected NCs for the same weight ratio was found to be only 78%. It was additionally found that the alignment of the NRs is mostly horizontal, since 82% are aligned within 10° of the x-y-plane.

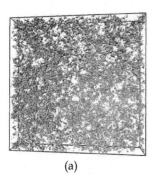   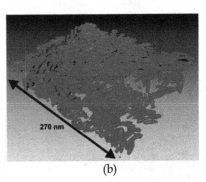

(a)   (b)

Fig. 10. (a) 3D visualizations of the a hybrid P3HT/ZnO hybrid film. Reprinted by permission from Macmillan Publishers Ltd: [Nature Materials] (Oosterhout et al., 2009), copyright (2009).; (b) Distribution of NRs within a OC_1C_{10}-PPV/CdSe-NR hybrid film based on TEM tomography. Reprinted with permission from (Hindson et al., 2011). Copyright 2011 American Chemical Society.

6.2.2 Morphology control by nanostructuring approaches

Morphology control on the nanoscale is a key issue to reduce the recombination of excitons. The optical absorption length within the donor material of the film is of about 100 nm (Peumans et al., 2003), while the generated excitons have a diffusion length of only 10 nm to 20 nm (Halls et al., 1996). Even if an exciton reaches the donor-acceptor interface before it recombines, the generated free charges must be extracted over continuous percolation pathways directly to the respective electrodes without being trapped or getting lost by charge recombination.

An interpenetrated donor acceptor structure on the nanoscale, as illustrated in Fig. 1d, would considerably improve the exciton diffusion, charge collection and charge transfer efficiency resulting in higher EQE value and so leading to a higher solar cell efficiency (Sagawa et al., 2010). Figure 1d is showing a conceptual design of an ideal structure of donor and acceptor phases within the heterojunction solar cell. Different nanostructuring approaches for hybrid heterojunction solar cells have been developed to implement such a device structure. A common method is the use of a porous template and the subsequent filling of the pores by a semiconducting material in order to fabricate vertically aligned nanopillars. One possibility to obtain porous templates is the anodic oxidation of Al to alumina, so-called Anodic Aluminum Oxidation (AAO) (Jessensky et al., 1998; Liu, P. A. et al., 2010). Here, vertical channels with diameters between 20 nm to 120 nm are formed by a first electrochemical oxidation and etching step, followed by a 2nd subsequent etching step for pore widening. The pores can be filled by different methods including simple pore filling, electrochemical deposition and vapor-liquid solid (VLS) growth processes. In principle the lengths, diameters and distances of the formed aligned nanopillars and nanowires can be controlled by the respective dimensions of the template and etching conditions. The height can be controlled by the thickness of the aluminium layer. In Fig. 11 a SEM image of an AAO template fabricated in our laboratory is shown. In a similar way the anodization of titanium films can lead to porous TiO_2 films and structures. The fabrication of vertically aligned tubes with pore diameters between 10 nm (Chen et al., 2007) and 100 nm (Macák et al., 2005) are reported. The main technical relevant differences to the AAO template is that TiO_2 itself is a semiconductor, while Al_2O_3 is an insulator, and that the pores in the TiO_2 template are closed at the bottom towards the ITO and so the filled in semiconducting material is not in contact with the electrode.

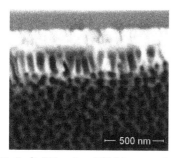 

Fig. 11. Left: Side view SEM image of a porous AAO membrane manufactured by anodic oxidation of aluminium at 40 V in the presence of 0.3 M oxalic acid; right: Side view SEM image of porous TiO_2 nanotubes fabricated by anodization of titanium. (Macák et al., 2005). Copyright Wiley-VCH Verlag GmbH & Co. KGaA. Reproduced with permission.

One notable example for the integration of a nanostructuring method into solar cell device fabrication is the use of AAO templates for the deposition of CdS by a VLS process leading to aligned nanopillars. Subsequent chemical vapor deposition (CVD) of CdTe resulted into a nanostructured all inorganic solar cell with an impressive PCE of ca. 6% (Fan et al., 2009). A few attempts to use vertically aligned nanopillars to obtain nanostructured hybrid solar cells also exist (Kuo et al., 2008; Ravirajan et al., 2006). These approaches resulted so far in devices with significant lower efficiencies compared to state of the art hybrid solar cells without additional nanostructuring steps. One example for the utilization of an AAO template for a nanostructured hybrid solar cell was published by Kuo et al. (Kuo et al., 2008) and is schematically illustrated in Fig. 12a together with its energy level diagram (Fig. 12b). A direct comparison between a nanostructured bulk-heterojunction hybrid solar cell and a bilayer based hybrid solar cell was performed. First, free standing nanopillars of TiO_2 were formed by spin coating of a TiO_2 dispersion onto the AAO template. After sintering at 450°C for 1 h and the subsequent removal of the 300 nm thick AAO template by NaOH, the TiO_2 nanopilllars were obtained. By covering the TiO_2 structure with P3HT via spin coating and subsequent evaporation of Au contacts, hybrid solar cells were manufactured with a PCE of 0.512% in comparison to 0.12% for the bilayer structure of the same donor-acceptor material composition. By this method an inverted solar cell was created, using gold as top electrode. A drawback in this design is that donor and acceptor materials are in direct contact with the ITO substrate, where both, holes and electrons, could be extracted leading to additional recombination events at the ITO electrode which lowers the overall solar cell efficiency.

(a)   (b)

Fig. 12. (a) Schematic illustration of an inverted TiO_2/P3HT hybrid solar cell manufactured by Kuo et al. using an AAO template for formation of parallel aligned TiO_2 nanopillars subsequently filled by P3HT; (b) Schematic illustration of the energy level diagram of the fabricated hybrid solar cell. Reprinted with permission from [Kuo et al., 2008]. Copyright [2008], American Institute of Physics.

It was demonstrated that by filling of the AAO template with a conjugated polymer, aligned polymer nanopillars were obtained exhibiting an increased hole mobility due to an improved vertical alignment of the polymeric chains within the AAO template (Coakley et al., 2005). The hole mobility rose by a factor of 20 from 3×10^{-4} $cm^2V^{-1}s^{-1}$ for a flat polymer layer in diode configuration to 6×10^{-3} $cm^2V^{-1}s^{-1}$ for the aligned polymer inside the AAO pores. After the AAO template was removed the spacings between the obtained polymer pillars can in principle be filled with an acceptor material like e.g. NCs from a deposited dispersion. This leads to a nanostructured hybrid solar cell with an interdigital device structure as illustrated in Fig. 1d.

Since TiO_2 is a semiconductor and could already be used as electron acceptor together with a conjugated donor polymer, the pores of a porous TiO_2 film could be directly filled with a donor polymer to obtain a nanostructured bulk-heterojunction hybrid film. Recently Lim et al. demonstrated the successful infiltration of P3HT into TiO_2 nanotubes of diameters of 60 nm to 80 nm. However, the diameters of the filled pores were above the desired diameters for an efficient charge extraction, so the reproducible and complete filling of the TiO_2 nanotubes is still one of the main challenges to be solved before this nanostructuring method can be implemented into hybrid solar cells.

Another method which was successfully applied for the formation of a nanostructured bulk-heterojunction organic solar cell is nanoimprint lithography (NIL). An AAO template was used as a mask for etching a Si substrate using a two-step inductively coupled plasma (ICP) etching process (Aryal et al., 2008). Thereby a silicon mold as shown in Fig. 13a is formed.

This mold is then used for creating NRs in a film of a conjugated polymer (e.g. regioregular P3HT). The created polymeric rods (Fig. 13c) show an increased crystallinity and preferential alignment of the polymer molecules in the vertical direction (Aryal et al., 2009) as well. The spacing between the polymer rods can then be filled with an acceptor material. After the evaporation of a top electrode the hybrid solar cell would be complete.

Kim et al. used NIL to create a nanostructured solar cell combining the molded polydithiophene derivative TDPDT with PCBM leading to a PCE of 0.8% compared to 0.25% of a bilayer structure (Kim et al., 2007).

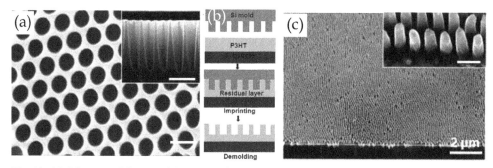

Fig. 13. (a) Silicon mold created by ICP etching using a AAO template as mask (inset image: side view of the mold); (b) illustration of the molding process applied to P3HT; (c) molded parallel aligned P3HT nanopillars. Reprinted with permission from (Aryal et al., 2009). Copyright 2009 American Chemical Society.

7. Outlook

Hybrid solar cells are still lagging behind the PCBM based OPV technology in respect of device performance and maturity for commercialization. They are currently under development and evaluation in basic research and have the potential for further significant improvement. The additional absorption of photons by semiconductor NCs, their potential to utilize multiple excitons generation and their higher electron conductivity compared to organic acceptor materials are some of the reasons behind. Novel device structures, the implementation of nanostructuring methods and the development of lower band gap material able to convert the NIR and IR parts of the solar spectrum into electrical energy will probably lead soon to PCE values of 10% and beyond for OPV technologies (Dennler et al., 2009). It is expected that the hybrid solar cell technologies also benefit from this development since device structure, nanostructuring methods and the development of novel low band gap polymers are overlapping aspects with pure OPV approaches. Progress in the development of organic-inorganic hybrid material design will not only be beneficial for the development of hybrid solar cells but also for various applications such as light emitting diodes, photodetectors etc. and have therefore a broader application potential beyond photovoltaics. In addition the energy levels in inorganic-organic hybrid materials can be tuned more easily compared to pure organic composites based on to the size quantization effects occurring in semiconductor nanostructures which might be beneficial for dedicated applications and allows a broad design flexibility for the variation of material composites.

Nevertheless one can clearly deduce from Table 1 that in all 1st and 2nd generation of PV technologies, differences between module PCEs and values of the best research cells are much smaller than in the case of DSSCs, OPV and hybrid solar cell technologies. Therefore the enhancement of the average module efficiencies of 3rd generation solar cells is one key issue to be addressed in order to extend this technology to wide range applications substituting traditional solar panels. In addition long-term stabilities of 3rd generation solar cells have to be improved tremendously to compete with existing PV technologies otherwise their utilization will be limited to small applications in devices with a limited lifetime such as e.g. disposable sensors and actuators. In case of hybrid solar cells the exploration of additional donor-acceptor materials is necessary, in order to replace toxic compounds by more environmental friendly materials.

8. Acknowledgment

Financial support from the German Federal Ministry of Education and Research (BMBF) within the project "NanoPolySol" under the contract No. 03X3517E as well as from the German Research Foundation (DFG) graduate school GRK 1322 "Micro Energy Harvesting" is gratefully acknowledged.

9. References

Aldakov, D.; Chandezon, F.; De Bettignies, R.; Firon, M.; Reiss, P. & Pron, A. (2006). Hybrid organic-inorganic nanomaterials: ligand effects. *European Physical Journal-Applied Physics*, Vol. 36, Nr. 3, pp. 261-265, ISSN 1286-0042

Alivisatos, A. P. (1996). Semiconductor clusters, nanocrystals, and quantum dots. *Science Science*, Vol. 271, Nr. 5251, pp. 933-937, ISSN 0036-8075

Arici, E.; Sariciftci, N. S. & Meissner, D. (2003). Hybrid solar cells based on nanoparticles of CuInS2 in organic matrices. *Adv Funct Mater*, Vol. 13, Nr. 2, pp. 165-171, ISSN 1616-301X

Aryal, M.; Buyukserin, F.; Mielczarek, K.; Zhao, X. M.; Gao, J. M.; Zakhidov, A. & Hu, W. C. (2008). Imprinted large-scale high density polymer nanopillars for organic solar cells. *Journal of Vacuum Science & Technology B*, Vol. 26, Nr. 6, pp. 2562-2566, ISSN 1071-1023

Aryal, M.; Trivedi, K. & Hu, W. C. (2009). Nano-Confinement Induced Chain Alignment in Ordered P3HT Nanostructures Defined by Nanoimprint Lithography. *Acs Nano*, Vol. 3, Nr. 10, pp. 3085-3090, ISSN 1936-0851

Babel, A. & Jenekhe, S. A. (2002). Electron transport in thin-film transistors from an n-type conjugated polymer. *Adv Mater Adv Mater*, Vol. 14, Nr. 5, pp. 371-374, ISSN 0935-9648

Beek, W. J. E.; Wienk, M. M. & Janssen, R. A. J. (2004). Efficient hybrid solar cells from zinc oxide nanoparticles and a conjugated polymer. *Adv Mater Adv Mater*, Vol. 16, Nr. 12, pp. 1009-1013, ISSN 0935-9648

Beek, W. J. E.; Wienk, M. M. & Janssen, R. A. J. (2006). Hybrid solar cells from regioregular polythiophene and ZnO nanoparticles. *Adv Funct Mater Adv Funct Mater*, Vol. 16, Nr. 8, pp. 1112-1116, ISSN 1616-301X

Benagli S.; Borrello D.; Vallat-Sauvain E.; Meier J.; Kroll U.; Hötzel J.; Spitznagel J.; Steinhauser J.; Castens L., & Djeridane Y. (2009). High-efficiency amorphous silicon devices on LPCVD-ZnO TCO prepared in industrial KAI-M R&D reactor. *24th European Photovoltaic Solar Energy Conference*, Hamburg, pp. 2293 - 2298

Borchert, H. (2010). Elementary processes and limiting factors in hybrid polymer/nanoparticle solar cells. *Energ Environ Sci* , Vol. 3, Nr. 11, pp. 1682-1694, ISSN 1754-5692

Brabec, C. J.; Winder, C.; Sariciftci, N. S.; Hummelen, J. C.; Dhanabalan, A.; van Hal, P. A. & Janssen, R. A. J. (2002). A low-bandgap semiconducting polymer for photovoltaic devices and infrared emitting diodes. *Adv Funct Mater Adv Funct Mater*, Vol. 12, Nr. 10, pp. 709-712, ISSN 1616-301X

Bredas, J. L.; Beljonne, D.; Coropceanu, V. & Cornil, J. (2004). Charge-transfer and energy-transfer processes in pi-conjugated oligomers and polymers: A molecular picture. *Chem Rev Chem Rev*, Vol. 104, Nr. 11, pp. 4971-5003, ISSN 0009-2665

Bruchez, M.; Moronne, M.; Gin, P.; Weiss, S. & Alivisatos, A. P. (1998). Semiconductor nanocrystals as fluorescent biological labels. *Science*, Vol. 281, Nr. 5385, pp. 2013-2016, ISSN 0036-8075

Brus, L. E. (1984). Electron Electron and Electron-Hole Interactions in Small Semiconductor Crystallites - the Size Dependence of the Lowest Excited Electronic State. *J Chem Phys*, Vol. 80, Nr. 9, pp. 4403-4409, ISSN 0021-9606

Chen, X.; Schriver, M.; Suen, T. & Mao, S. S. (2007). Fabrication of 10 nm diameter TiO2 nanotube arrays by titanium anodization. *Thin Solid Films*, Vol. 515, Nr. 24, pp. 8511-8514, ISSN 0040-6090

Coakley, K. M.; Srinivasan, B. S.; Ziebarth, J. M.; Goh, C.; Liu, Y. X. & McGehee, M. D. (2005). Enhanced hole mobility in regioregular polythiophene infiltrated in straight nanopores. *Advanced Functional Materials*, Vol. 15, Nr. 12, pp. 1927-1932, ISSN 1616-301X

Cui, D. H.; Xu, J.; Zhu, T.; Paradee, G.; Ashok, S. & Gerhold, M. (2006). Harvest of near infrared light in PbSe nanocrystal-polymer hybrid photovoltaic cells. *Appl Phys Lett Appl Phys Lett*, Vol. 88, Nr. 18, pp. 183111, ISSN 0003-6951

Cunningham, D.; Davies, K.; Grammond, L.; Mopas, E.; O Connor, N.; Rubchich, M.; Sadeghi, M.; Skinner, N., & Trumbly, T. (2000). *28th IEEE Photovoltaic Specialists Conference*, Alaska, USA, pp. 13-18

Dayal, S.; Kopidakis, N.; Olson, D. C.; Ginley, D. S. & Rumbles, G. (2010). Photovoltaic Devices with a Low Band Gap Polymer and CdSe Nanostructures Exceeding 3% Efficiency. *Nano Lett Nano Lett*, Vol. 10, Nr. 1, pp. 239-242, ISSN 1530-6984

Dayal, S.; Reese, M. O.; Ferguson, A. J.; Ginley, D. S.; Rumbles, G. & Kopidakis, N. (2010). The Effect of Nanoparticle Shape on the Photocarrier Dynamics and Photovoltaic Device Performance of Poly(3-hexylthiophene):CdSe Nanoparticle Bulk Heterojunction Solar Cells., pp. 3629-2635

Dennler, G.; Scharber, M. C. & Brabec, C. J. (2009). Polymer-Fullerene Bulk-Heterojunction Solar Cells. *Adv Mater Adv Mater*, Vol. 21, Nr. 13, pp. 1323-1338, ISSN 0935-9648

Diener, M. D. & Alford, J. M. (1998). Isolation and properties of small-bandgap fullerenes. *Nature Nature*, Vol. 393, Nr. 6686, pp. 668-671, ISSN 0028-0836

Erb, T.; Zhokhavets, U.; Gobsch, G.; Raleva, S.; Stuhn, B.; Schilinsky, P.; Waldauf, C. & Brabec, C. J. (2005). Correlation between structural and optical properties of composite polymer/fullerene films for organic solar cells. *Advanced Functional Materials*, Vol. 15, Nr. 7, pp. 1193-1196, ISSN 1616-301X

Fan, Z. Y.; Razavi, H.; Do, J. W.; Moriwaki, A.; Ergen, O.; Chueh, Y. L.; Leu, P. W.; Ho, J. C.; Takahashi, T.; Reichertz, L. A.; Neale, S.; Yu, K.; Wu, M.; Ager, J. W. & Javey, A.

(2009). Three-dimensional nanopillar-array photovoltaics on low-cost and flexible substrates. *Nature Materials*, Vol. 8, Nr. 8, pp. 648-653, ISSN 1476-1122

Frechet, J. M. J.; Holcombe, T. W.; Woo, C. H.; Kavulak, D. F. J. & Thompson, B. C. (2009). All-Polymer Photovoltaic Devices of Poly(3-(4-n-octyl)-phenylthiophene) from Grignard Metathesis (GRIM) Polymerization. *J Am Chem Soc J Am Chem Soc*, Vol. 131, Nr. 40, pp. 14160-14161, ISSN 0002-7863

Ginger, D. S. & Greenham, N. C. (2000). Charge injection and transport in films of CdSe nanocrystals. *J Appl Phys J Appl Phys*, Vol. 87, Nr. 3, pp. 1361-1368, ISSN 0021-8979

Gledhill, S. E.; Scott, B. & Gregg, B. A. (2005). Organic and nano-structured composite photovoltaics: An overview. *J Mater Res J Mater Res*, Vol. 20, Nr. 12, pp. 3167-3179, ISSN 0884-2914

Goldstein, J.; Yakupov, I. & Breen, B. (2010). Development of large area photovoltaic dye cells at 3GSolar. *Solar Energy Materials and Solar Cells*, Vol. 94, Nr. 4, pp. 638-641, ISSN 0927-0248

Greenham, N. C. (2008) Hybrid Polymer/Nanocrystal Photovoltaic Devices, in Organic Photovoltaics (eds C. Brabec, V. Dyakonov and U. Scherf), Wiley-VCH Verlag GmbH & Co. KGaA, Weinheim, Germany.

Greenham, N. C.; Peng, X. G. & Alivisatos, A. P. (1996). Charge separation and transport in conjugated-polymer/semiconductor-nanocrystal composites studied by photoluminescence quenching and photoconductivity. *Phys Rev B Phys Rev B*, Vol. 54, Nr. 24, pp. 17628-17637, ISSN 1098-0121

Gunes, S.; Fritz, K. P.; Neugebauer, H.; Sariciftci, N. S.; Kumar, S. & Scholes, G. D. (2007). Hybrid solar cells using PbS nanoparticles. *Sol Energ Mat Sol C Sol Energ Mat Sol C*, Vol. 91, Nr. 5, pp. 420-423, ISSN 0927-0248

Gur, I.; Fromer, N. A.; Chen, C. P.; Kanaras, A. G. & Alivisatos, A. P. (2007). Hybrid solar cells with prescribed nanoscale morphologies based on hyperbranched semiconductor nanocrystals. *Nano Lett Nano Lett*, Vol. 7, Nr. 2, pp. 409-414, ISSN 1530-6984

Halls, J. J. M.; Pichler, K.; Friend, R. H.; Moratti, S. C. & Holmes, A. B. (1996). Exciton diffusion and dissociation in a poly(p-phenylenevinylene)/C-60 heterojunction photovoltaic cell. *Appl Phys Lett Appl Phys Lett*, Vol. 68, Nr. 22, pp. 3120-3122, ISSN 0003-6951

Halls, J. J. M.; Walsh, C. A.; Greenham, N. C.; Marseglia, E. A.; Friend, R. H.; Moratti, S. C. & Holmes, A. B. (1995). Efficient Photodiodes from Interpenetrating Polymer Networks. *Nature Nature*, Vol. 376, Nr. 6540, pp. 498-500, ISSN 0028-0836

Han, L.; Fukui, A.; Fuke, N.; Koide, N., & Yamanaka, R. (2006). *4th World Conference on Photovoltaic Energy Conversion (WCEP-4)*, Hawai, USA

Heliatek, Heliatek and IAPP achieve production-relevant efficiency record for organic photovoltaic cells, (11-10-2010), available at: http://www.heliatek.com/news-19

Hindson, J. C.; Saghi, Z.; Hernandez-Garrido, J. C.; Midgley, P. A. & Greenham, N. C. (2011). Morphological Study of Nanoparticle-Polymer Solar Cells Using High-Angle Annular Dark-Field Electron Tomography. *Nano Letters*, Vol. 11, Nr. 2, pp. 904-909, ISSN 1530-6984

Huynh, W. U.; Dittmer, J. J. & Alivisatos, A. P. (2002). Hybrid nanorod-polymer solar cells. *Science Science*, Vol. 295, Nr. 5564, pp. 2425-2427, ISSN 0036-8075

Huynh, W. U.; Dittmer, J. J.; Libby, W. C.; Whiting, G. L. & Alivisatos, A. P. (2003). Controlling the morphology of nanocrystal-polymer composites for solar cells. *Advanced Functional Materials*, Vol. 13, Nr. 1, pp. 73-79, ISSN 1616-301X

Huynh, W. U.; Dittmer, J. J.; Teclemariam, N.; Milliron, D. J.; Alivisatos, A. P. & Barnham, K. W. J. (2003). Charge transport in hybrid nanorod-polymer composite photovoltaic cells. *Phys Rev B Phys Rev B* , Vol. 67, Nr. 11, pp. 115326, ISSN 1098-0121

Jackson P.; Hariskos D.; Lotter E.; Paetel S.; Wuerz R.; Menner R.; Wischmann W. & Powalla M., (2011). New world record efficiency for Cu(In,Ga)Se2 thin-film solar cells beyond 20%. *Progress in Photovoltaics: Research and Applications*. Published online. DOI: 10.1002/pip.1078

Jessensky, O.; Muller, F. & Gosele, U. (1998). Self-organized formation of hexagonal pore arrays in anodic alumina. *Applied Physics Letters*, Vol. 72, Nr. 10, pp. 1173-1175, ISSN 0003-6951

Kamat, P. V. (2008). Quantum Dot Solar Cells. Semiconductor Nanocrystals as Light Harvesters. *J Phys Chem C J Phys Chem C*, Vol. 112, Nr. 48, pp. 18737-18753, ISSN 1932-7447

Kang, Y. M.; Park, N. G. & Kim, D. (2005). Hybrid solar cells with vertically aligned CdTe nanorods and a conjugated polymer. *Appl Phys Lett Appl Phys Lett*, Vol. 86, Nr. 11, ISSN 0003-6951

Kazes, M.; Lewis, D. Y.; Ebenstein, Y.; Mokari, T. & Banin, U. (2002). Lasing from semiconductor quantum rods in a cylindrical microcavity. *Adv Mater Adv Mater*, Vol. 14, Nr. 4, pp. 317, ISSN 0935-9648

Kietzke, T. (2007). Recent Advances in Organic Solar Cells., Vol. 2007, pp. 40285

Kim, M. S.; Kim, J. S.; Cho, J. C.; Shtein, M.; Guo, L. J. & Kim, J. (2007). Flexible conjugated polymer photovoltaic cells with controlled heterojunctions fabricated using nanoimprint lithography. *Applied Physics Letters*, Vol. 90, Nr. 12, ISSN 0003-6951

Konarka Technologies, Konarka's Power Plastic Achieves World Record 8.3% Efficiency Certification from National Energy Renewable Laboratory (NREL), (29-11-2010), available at: http://www.konarka.com/index.php/site/pressreleasedetail/konarkas_power_plastic_achieves_world_record_83_efficiency_certification_fr

Kumar, S. & Nann, T. (2004). First solar cells based on CdTe nanoparticle/MEH-PPV composites. *J Mater Res J Mater Res*, Vol. 19, Nr. 7, pp. 1990-1994, ISSN 0884-2914

Kuo, C. Y.; Tang, W. C.; Gau, C.; Guo, T. F. & Jeng, D. Z. (2008). Ordered bulk heterojunction solar cells with vertically aligned TiO2 nanorods embedded in a conjugated polymer. *Applied Physics Letters*, Vol. 93, Nr. 3, ISSN 0003-6951

Liao, H. C.; Chen, S. Y. & Liu, D. M. (2009). In-Situ Growing CdS Single-Crystal Nanorods via P3HT Polymer as a Soft Template, for Enhancing Photovoltaic Performance. *Macromolecules*, Vol. 42, Nr. 17, pp. 6558-6563, ISSN 0024-9297

Lim, S. L.; Liu, Y. L.; Liu, G.; Xu, S. Y.; Pan, H. Y.; Kang, E. T. & Ong, C. K. (2011). Infiltrating P3HT polymer into ordered TiO2 nanotube arrays. *Physica Status Solidi A-Applications and Materials Science*, Vol. 208, Nr. 3, pp. 658-663, ISSN 1862-6300

Liu, C. Y.; Holman, Z. C. & Kortshagen, U. R. (2009). Hybrid Solar Cells from P3HT and Silicon Nanocrystals. *Nano Lett Nano Lett*, Vol. 9, Nr. 1, pp. 449-452, ISSN 1530-6984

Liu, C. Y.; Holman, Z. C. & Kortshagen, U. R. (2010). Optimization of Si NC/P3HT Hybrid Solar Cells., Vol. 20, Nr. 13, pp. 2157-2164

Liu, P. A.; Singh, V. P. & Rajaputra, S. (2010). Barrier layer non-uniformity effects in anodized aluminum oxide nanopores on ITO substrates. *Nanotechnology*, Vol. 21, Nr. 11, ISSN 0957-4484

Macak, J. M.; Tsuchiya, H. & Schmuki, P. (2005). High-aspect-ratio TiO2 nanotubes by anodization of titanium. *Angewandte Chemie-International Edition*, Vol. 44, Nr. 14, pp. 2100-2102, ISSN 1433-7851

MiaSolé, MiaSolé Achieves 15.7% Efficiency with Commercial-Scale CIGS Thin Film Solar Modules, (2-12-2010), available at: http://www.miasole.com/sites/ default/files/MiaSole_release_Dec_02_2010. pdf

Mitsubishi Chemical, 8.5% efficient small molecule organic solar cell, (8-3-2011), available at: http://www.physorg.com/pdf218812262.pdf

Morana, M.; Wegscheider, M.; Bonanni, A.; Kopidakis, N.; Shaheen, S.; Scharber, M.; Zhu, Z.; Waller, D.; Gaudiana, R. & Brabec, C. (2008). Bipolar charge transport in PCPDTBT-PCBM bulk-heterojunctions for photovoltaic applications. *Adv Funct Mater Adv Funct Mater*, Vol. 18, Nr. 12, pp. 1757-1766, ISSN 1616-301X

Musselman, K. P.; Mulholland, G. J.; Robinson, A. P.; Schmidt-Mende, L. & MacManus-Driscoll, J. L. (2008). Low-Temperature Synthesis of Large-Area, Free-Standing Nanorod Arrays on ITO/Glass and other Conducting Substrates. *Advanced Materials*, Vol. 20, Nr. 23, pp. 4470-4475, ISSN 0935-9648

Musselman, K. P.; Wisnet, A.; Iza, D. C.; Hesse, H. C.; Scheu, C.; MacManus-Driscoll, J. L. & Schmidt-Mende, L. (2010). Strong Efficiency Improvements in Ultra-low-Cost Inorganic Nanowire Solar Cells. *Advanced Materials*, Vol. 22, Nr. 35, pp. E254-E258, ISSN 0935-9648

Niggemann, M.; Glatthaar, M.; Gombert, A.; Hinsch, A. & Wittwer, V. (2004). Diffraction gratings and buried nano-electrodes - architectures for organic solar cells. *Thin Solid Films*, Vol. 451-52, pp. 619-623, ISSN 0040-6090

Niggemann, M.; Riede, M.; Gombert, A. & Leo, K. (2008). Light trapping in organic solar cells. *Physica Status Solidi A-Applications and Materials Science*, Vol. 205, Nr. 12, pp. 2862-2874, ISSN 1862-6300

Olson, J. D.; Gray, G. P. & Carter, S. A. (2009). Optimizing hybrid photovoltaics through annealing and ligand choice. *Sol Energ Mat Sol C Sol Energ Mat Sol C*, Vol. 93, Nr. 4, pp. 519-523, ISSN 0927-0248

Oosterhout, S. D.; Wienk, M. M.; van Bavel, S. S.; Thiedmann, R.; Koster, L. J. A.; Gilot, J.; Loos, J.; Schmidt, V. & Janssen, R. A. J. (2009). The effect of three-dimensional morphology on the efficiency of hybrid polymer solar cells. *Nat Mater Nat Mater*, Vol. 8, Nr. 10, pp. 818-824, ISSN 1476-1122

Owen, J. S.; Park, J.; Trudeau, P. E. & Alivisatos, A. P. (2008). Reaction chemistry and ligand exchange at cadmium-selenide nanocrystal surfaces. *J Am Chem Soc J Am Chem Soc*, Vol. 130, Nr. 37, pp. 12279-12281, ISSN 0002-7863

Panthani, M. G.; Akhavan, V.; Goodfellow, B.; Schmidtke, J. P.; Dunn, L.; Dodabalapur, A.; Barbara, P. F. & Korgel, B. A. (2008). Synthesis of CuInS2, CuInSe2, and Cu(InxGa1-x)Se-2 (CIGS) Nanocrystal "Inks" for Printable Photovoltaics. *Journal of the American Chemical Society*, Vol. 130, Nr. 49, pp. 16770-16777, ISSN 0002-7863

Park, S. H.; Roy, A.; Beaupre, S.; Cho, S.; Coates, N.; Moon, J. S.; Moses, D.; Leclerc, M.; Lee, K. & Heeger, A. J. (2009). Bulk heterojunction solar cells with internal quantum efficiency approaching 100%. *Nat Photonics Nat Photonics*, Vol. 3, Nr. 5, pp. 297-302, ISSN 1749-4885

Peet, J.; Kim, J. Y.; Coates, N. E.; Ma, W. L.; Moses, D.; Heeger, A. J. & Bazan, G. C. (2007). Efficiency enhancement in low-bandgap polymer solar cells by processing with alkane dithiols. *Nature Materials*, Vol. 6, Nr. 7, pp. 497-500, ISSN 1476-1122

Peng, X. G.; Manna, L.; Yang, W. D.; Wickham, J.; Scher, E.; Kadavanich, A. & Alivisatos, A. P. (2000). Shape control of CdSe nanocrystals. *Nature Nature*, Vol. 404, Nr. 6773, pp. 59-61, ISSN 0028-0836

Peng, Z. A. & Peng, X. G. (2001). Formation of high-quality CdTe, CdSe, and CdS nanocrystals using CdO as precursor. *J Am Chem Soc J Am Chem Soc*, Vol. 123, Nr. 1, pp. 183-184, ISSN 0002-7863

Peumans, P.; Yakimov, A. & Forrest, S. R. (2003). Small molecular weight organic thin-film photodetectors and solar cells. *J Appl Phys J Appl Phys*, Vol. 93, Nr. 7, pp. 3693-3723, ISSN 0021-8979

Ravirajan, P.; Peiro, A. M.; Nazeeruddin, M. K.; Graetzel, M.; Bradley, D. D. C.; Durrant, J. R. & Nelson, J. (2006). Hybrid polymer/zinc oxide photovoltaic devices with vertically oriented ZnO nanorods and an amphiphilic molecular interface layer. *Journal of Physical Chemistry B*, Vol. 110, Nr. 15, pp. 7635-7639, ISSN 1520-6106

Riede, M.; Mueller, T.; Tress, W.; Schueppel, R. & Leo, K. (2008). Small-molecule solar cells - status and perspectives. *Nanotechnology Nanotechnology*, Vol. 19, Nr. 42, pp. 424001, ISSN 0957-4484

Rim, S. B.; Zhao, S.; Scully, S. R.; McGehee, M. D. & Peumans, P. (2007). An effective light trapping configuration for thin-film solar cells. *Applied Physics Letters*, Vol. 91, Nr. 24, ISSN 0003-6951

Rode, D. L. (1970). Electron mobility in II-VI semiconductors. *Phys Rev B-Solid St Phys Rev B-Solid St*, Vol. 2, Nr. 10, pp. 4036-4044

Roman, L. S.; Inganas, O.; Granlund, T.; Nyberg, T.; Svensson, M.; Andersson, M. R. & Hummelen, J. C. (2000). Trapping light in polymer photodiodes with soft embossed gratings. *Advanced Materials*, Vol. 12, Nr. 3, pp. 189-+, ISSN 0935-9648

Sagawa, T.; Yoshikawa, S. & Imahori, H. (2010). One-Dimensional Nanostructured Semiconducting Materials for Organic Photovoltaics. *Journal of Physical Chemistry Letters*, Vol. 1, Nr. 7, pp. 1020-1025, ISSN 1948-7185

Saunders, B. R. & Turner, M. L. (2008). Nanoparticle-polymer photovoltaic cells. *Advances in Colloid and Interface Science*, Vol. 138, Nr. 1, pp. 1-23, ISSN 0001-8686

Scharber, M. C.; Wuhlbacher, D.; Koppe, M.; Denk, P.; Waldauf, C.; Heeger, A. J. & Brabec, C. L. (2006). Design rules for donors in bulk-heterojunction solar cells - Towards 10 % energy-conversion efficiency. *Adv Mater Adv Mater*, Vol. 18, Nr. 6, pp. 789-794, ISSN 0935-9648

Schilinsky, P.; Waldauf, C. & Brabec, C. J. (2002). Recombination and loss analysis in polythiophene based bulk heterojunction photodetectors. *Appl Phys Lett Appl Phys Lett*, Vol. 81, Nr. 20, pp. 3885-3887, ISSN 0003-6951

Schott Solar, SCHOTT Solar Presents Champion Multicrystalline Module, (7-9-2010), available at: http://www.schott.com/english/news/press.html?NID=2948

Schultz, O.; Glunz, S. W. & Willeke, G. P. (2004). Multicrystalline silicon solar cells exceeding 20% efficiency. *Progress in Photovoltaics*, Vol. 12, Nr. 7, pp. 553-558, ISSN 1062-7995

Seo, J.; Kim, W. J.; Kim, S. J.; Lee, K. S.; Cartwright, A. N. & Prasad, P. N. (2009). Polymer nanocomposite photovoltaics utilizing CdSe nanocrystals capped with a thermally cleavable solubilizing ligand. *Appl Phys Lett Appl Phys Lett*, Vol. 94, Nr. 13, pp. 133302, ISSN 0003-6951

Sih, B. C. & Wolf, M. (2007). CdSe nanorods functionalized with thiol-anchored oligothiophenes. *J Phys Chem C J Phys Chem C*, Vol. 111, Nr. 46, pp. 17184-17192, ISSN 1932-7447

Soci, C.; Hwang, I. W.; Moses, D.; Zhu, Z.; Waller, D.; Gaudiana, R.; Brabec, C. J. & Heeger, A. J. (2007). Photoconductivity of a low-bandgap conjugated polymer. *Adv Funct Mater Adv Funct Mater*, Vol. 17, Nr. 4, pp. 632-636, ISSN 1616-301X

Solarmer Energy, Press release, (18-6-2009), available at: http://www.printedelectronicsnow.com/news/2009/06/23/solamer_energy_ picks_up_speed_in_flexible_solar_panel_development_

Su, Z. X. & Zhou, W. Z. (2008). Formation Mechanism of Porous Anodic Aluminium and Titanium Oxides. *Advanced Materials*, Vol. 20, Nr. 19, pp. 3663-3667, ISSN 0935-9648

Sun, B. Q. & Greenham, N. C. (2006). Improved effciency of photovoltaics based on CdSe nanorods and poly(3-hexylthiophene) nanofibers. *Physical Chemistry Chemical Physics*, Vol. 8, Nr. 30, pp. 3557-3560, ISSN 1463-9076

Sun, B. Q.; Snaith, H. J.; Dhoot, A. S.; Westenhoff, S. & Greenham, N. C. (2005). Vertically segregated hybrid blends for photovoltaic devices with improved efficiency. *J Appl Phys J Appl Phys*, Vol. 97, Nr. 1, pp. 014914, ISSN 0021-8979

van Beek, R.; Zoombelt, A. P.; Jenneskens, L. W.; van Walree, C. A.; Donega, C. D.; Veldman, D. & Janssen, R. A. J. (2006). Side chain mediated electronic contact between a tetrahydro-4H-thiopyran-4-ylidene-appended polythiophene and CdTe quantum dots. *Chem-Eur J Chem-Eur J*, Vol. 12, Nr. 31, pp. 8075-8083, ISSN 0947-6539

Wang, P.; Abrusci, A.; Wong, H. M. P.; Svensson, M.; Andersson, M. R. & Greenham, N. C. (2006). Photoinduced charge transfer and efficient solar energy conversion in a blend of a red polyfluorene copolymer with CdSe nanoparticles. *Nano Lett Nano Lett*, Vol. 6, Nr. 8, pp. 1789-1793, ISSN 1530-6984

Watt, A. A. R.; Blake, D.; Warner, J. H.; Thomsen, E. A.; Tavenner, E. L.; Rubinsztein-Dunlop, H. & Meredith, P. (2005). Lead sulfide nanocrystal: conducting polymer solar cells. *Journal of Physics D-Applied Physics*, Vol. 38, Nr. 12, pp. 2006-2012, ISSN 0022-3727

Watt, A.; Thomsen, E.; Meredith, P. & Rubinsztein-Dunlop, H. (2004). A new approach to the synthesis of conjugated polymer-nanocrystal composites for heterojunction optoelectronics. *Chemical Communications*, Nr. 20, pp. 2334-2335, ISSN 1359-7345

Wu X.; Keane J.C.; Dhere R.G.; DeHart C.; Duda A.; Asher S.; Levi D.H., & Sheldon P. (2001). 16.5%-efficient CdS/CdTe polycrystalline thin-film solar cell. *17th European Photovoltaic Solar Energy Conference*, München, pp. 995-1000

Wu, Y. & Zhang, G. (2010). Performance Enhancement of Hybrid Solar Cells Through Chemical Vapor Annealing., Vol. 10, pp. 1628-1631

Xu, T. & Qiao, Q. (2011). Conjugated polymer-inorganic semiconductor hybrid solar cells. *Energy & Environmental Science*, DOI: 10.1039/c0ee00632g

Yue, W. J.; Han, S. K.; Peng, R. X.; Shen, W.; Geng, H. W.; Wu, F.; Tao, S. W. & Wang, M. T. (2010). CuInS2 quantum dots synthesized by a solvothermal route and their application as effective electron acceptors for hybrid solar cells. *Journal of Materials Chemistry*, Vol. 20, Nr. 35, pp. 7570-7578, ISSN 0959-9428

Zhao, J. H.; Wang, A. H.; Green, M. A. & Ferrazza, F. (1998). 19.8% efficient "honeycomb" textured multicrystalline and 24.4% monocrystalline silicon solar cells. *Applied Physics Letters*, Vol. 73, Nr. 14, pp. 1991-1993, ISSN 0003-6951

Zhao, J.; Wang, A.; Yun, F.; Zhang, G.; Roche, D. M.; Wenham, S. R. & Green, M. A. (1997). 20,000 PERL silicon cells for the '1996 world solar challenge' solar car race. *Progress in Photovoltaics*, Vol. 5, Nr. 4, pp. 269-276, ISSN 1062-7995

Zhou, Y. F.; Eck, M. & Krüger, M. (2010). Bulk-heterojunction hybrid solar cells based on colloidal nanocrystals and conjugated polymers. *Energy & Environmental Science*, Vol. 3, Nr. 12, pp. 1851-1864, ISSN 1754-5692

Zhou, Y. F.; Riehle, F. S.; Yuan, Y.; Schleiermacher, H. F.; Niggemann, M.; Urban, G. A. & Krueger, M. (2010). Improved efficiency of hybrid solar cells based on non-ligand-exchanged CdSe quantum dots and poly(3-hexylthiophene). *Appl Phys Lett Appl Phys Lett*, Vol. 96, Nr. 1, pp. 013304, ISSN 0003-6951

Zhou, Y.; Eck, M.; Veit, C.; Zimmermann, B.; Rauscher, F.; Niyamakom, P.; Yilmaz, S.; Dumsch, I.; Allard, S.; Scherf, U. & Krueger, M. (2011). Efficiency enhancement for bulk-heterojunction hybrid solar cells based on acid treated CdSe quantum dots and low bandgap polymer PCPDTBT. *Solar Energy Materials and Solar Cells*, Vol. 95, Nr. 4, pp. 1232-1237, ISSN 0927-0248

Zhou, Y.; Li, Y. C.; Zhong, H. Z.; Hou, J. H.; Ding, Y. Q.; Yang, C. H. & Li, Y. F. (2006). Hybrid nanocrystal/polymer solar cells based on tetrapod-shaped CdSexTe1-x nanocrystals. *Nanotechnology Nanotechnology*, Vol. 17, Nr. 16, pp. 4041-4047, ISSN 0957-4484

5

Dilute Nitride GaAsN and InGaAsN Layers Grown by Low-Temperature Liquid-Phase Epitaxy

Malina Milanova[1] and Petko Vitanov[2]
[1]*Central Laboratory of Applied Physics, BAS*
[2]*Central Laboratory of New Energy & New Energy Sources, BAS*
Bulgaria

1. Introduction

A critical goal for photovoltaic energy conversion is the development of high-efficiency, low cost photovoltaic structures which can reach the thermodynamic limit of solar energy conversion. New concepts aim to make better use of the solar spectrum than conventional single-gap cells currently do. In multijunction solar cells based on III-V heterostructures, better spectrum utilization is obtained by stacking several solar cells. These cells have achieved the highest efficiency among all other solar cells and have the theoretical potential to achieve efficiencies equivalent to or exceeding all other approaches. Record conversion efficiencies of 40.7 % (King, 2008) and 41,1 (Guter at al., 2009) under concentrated light for triple- junction allows hoping for practical realization of gianed values of efficiency in more multiplejunction structures. The expectations will be met , if suitable novel materials for intermediate cascades are found, and these materials are grown of an appropriate quality. Models indicate that higher efficiency would be obtained for 4-junction cells where 1.0 eV band gap cell is added in series to proven InGaP/GaAs/Ge triple-junction structures. Dilute nitride alloys such as GaInAsN, GaAsSbN provide a powerful tool for engineering the band gap and lattice constant of III-V alloys, due to their unique properties. They are promising novel materials for 4- and 5-junction solar cells performance. They exhibit strong bowing parameters and hold great potential to extend the wavelength further to the infrared part of the spectrum.

The incorporation of small quantity of nitrogen into GaAs causes a dramatic reduction of the band gap (Weyeres et al., 1992), but it also deteriorates the crystalline and optoelectronic properties of the dilute nitride materials, including reduction of the photoluminescence intensity and lifetime, reduction of electron mobility and increase in the background carrier concentration. Technologically, the incorporation probability of nitrogen in GaAs is very small and strongly depends on the growth conditions. GaAsN- based alloys and heterostructures are primarily grown by metaloorganic vapor-phase epitaxy (MOVPE) (Kurtz et all, 2000; Johnston et all, 2005)) and molecular-beam epitaxy (MBE) (Kurtz et al. 2002; Krispin et al, 2002; Khan et al, 2007), but the material quality has been inferior to that of GaAs. A peak internal quantum efficiency of 70 % is obtained for the solar cells grown by MOCVD (Kurtz et al. 1999). Internal quantum values near to unit are reported for p-i-n

GaInAsN cell grown by MBE (Ptak et al 2005), but photovoltages in this material are still low. Recently chemical-beam epitaxy (Nishimura et al., 2007; Yamaguchi et al, 2008; Oshita et al, 2011) has been developed in order to improve the quality of the grown layer, but today it remains a challenge to grow dilute nitride materials with photovoltaic (PV) quality.

In this chapter we present some results on thick GaAsN and InGaAsN layers, grown by low-temperature Liquid-Phase Epitaxy (LPE). In the literature there are only a few works on dilute nitride GaAsN grown by LPE (Dhar et al., 2005; Milanova et al., 2009) and some data for InGaAsN (Vitanov et al., 2010).

2. Heteroepitaxy nucleation and growth modes

The mechanism of nucleation and initial growth stage of heteroepitaxy dependence on bonding between the layer and substrate across the interface. Since the heteroepitaxy requires the nucleation of a new alloy on a foreign substrate the surface chemistry and physics play important roles in determining the properties of heteroepitaxial growth. In the classical theory, the mechanism of heterogeneous nucleation is determined by the surface and interfacial free energies for the substrate and epitaxial crystal.

Three classical modes of initial growth introduced at first by Ernst Bauer in 1958 can be distinguished: Layer by layer or Frank–Van der Merwe FM two-dimension mode (Frank–Van der Merwe, 1949), Volmer–Weber VW 3D island mode (Volmer–Weber, 1926), and Stranski-Krastanov SK or layer-plus-island mode (Stranski-Krastanov, 1938) as the intermediate case. The layer by layer growth mode arises when dominates the interfacial energy between substrate and epilayer material. In the opposite case, for the weak interfacial energy when the deposit atoms are more strongly bound to each other than they are to the substrate, the island (3D), or VW mode results. In the SK case, 3D island are formed on several monolayers, grown in a layer-by-layer on a crystal substrate.

Schematically these growth modes are shown in the Figure 2.1.

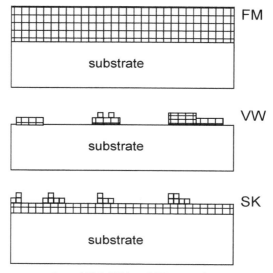

Fig. 2.1. Schematic presentation of FM, VW and SK growth modes

The growth modes in heteroepitaxy are defined based on thermodynamic models.

The sum of the film surface energy and the interface energy must be less than the surface energy of the substrate in order for wetting to occur and then layer by layer growth is expected. The VW growth mode is to be expected for a no wetting epitaxial layer. If γ and γ_0 are the surface free energies of the layer and substrate, respectively, and γ_i is the interfacial free energy the change in the free energy $\Delta\gamma$ associated with covering the substrate with epitaxial layer is:

$$\Delta\gamma = \gamma + \gamma_i - \gamma_0 \qquad (2.1)$$

If minimum energy determinates the mode for nucleation and growth, the dominated mechanism will be two-dimensional for $\Delta\gamma < 0$ and three-dimensional for $\Delta\gamma > 0$. However, even in the case of a wetting epitaxial layer ($\Delta\gamma < 0$), the existence of mismatch strain can cause islanding after the growth of a few monolayers. This is because the strain energy , increases linearly with the number of strained layers. At some thickness, $\gamma + \gamma_i$ exceeds γ_0 and the growth mode transforms from FM to SK resulting in 3D islands on the 2D wetting layer. Whereas it is clear that the VW growth mode is expected for a nonwetting epitaxial layer, the behavior of a wetting deposit is more complex and requires further consideration. Often the interfacial contribution in the limit of zero lattice mismatch and weak chemical interactions between the film and substrate at the interface can be neglected in comparison to the surface free energy ($\gamma_i \approx 0$). In this case the growth mode is determined entirely by the surface free energies of the film and substrate material.

Instead of these three main growth modes additional growth modes and epitaxial growth mechanisms could be distinguished (Scheel, 2003): columnar growth, step flow mode, step bunching, and screw-island growth.

The structural quality of the layer and surface morphology strongly depend on the growth method and the main growth parameters: supersaturation, misorientation of the substrate and the difference of lattice constants between substrate and the epitaxial layer.

In the case of flat substrate, the supersaturation increases until surface nucleation of a new monolayer occurs and its growth cover the substrate, followed by the nucleation of the next monolayer. For compound of limited thermodynamic stability or with volatile constituents like GaAs, GaN, SiC the appearance of the growth mode is largely predetermined by the choice of the growth method due to the inherent high supersaturation in epitaxy from the vapor phase and adjustable low supersaturation in LPE.

The FM growth mode in LPE can only be obtained at quasi-zero misfit as it is established from thermodynamic theory (Van der Merwe, 1979) and demonstrated by atomistic simulations using the Lennard–Jones potential (Grabow and Gilmer, 1988) and also at low supersaturation. At high supersaturation a high thermodynamic driving force leads to a high density of steps moving with large step velocities over the surface and causes step bunching.

The VW mode is typical of VPE. Due to the high supersaturation a large number of surface nuclei arise, which then spread and form three-dimensional islands, that finally coalesce to a compact layer. Continued growth of a layer initiated by the VW mode often shows columnar growth which is a common feature in epitaxy of GaN and diamond. (Hiramatsu *et al.*, 1991). The SK mode has been demonstrated by MBE growth of InAs onto GaAs substrate (Nabetani *et al.*, 1994).

Observations, analyses and measurements of LPE GaAs on the formation of nuclei and surface terraces show that nuclei grow into well-defined prismatic hillocks bounded by only {100} and {111} planes and they are unique to each substrate orientation, and hillocks tend to coalesce into chains and then into parallel surface terraces (Mattes & Route, 1974). The hillock boundaries may cause local strain fields and variation of the incorporation rates of impurities and dopants, or the local strain may getter or rejects impurities during annealing processes. This inhomogeneity may be suppressed by providing one single step source or by using substrates of well-defined small misorientation. The FM growth mode and such homogeneous layers can only be achieved by LPE or by VPE at very high growth temperatures.

Only at low supersaturation, nearly zero misfit and small misorientation of the substrate the layer by-layer growth mode can be realized and used to produce low dislocation layers for ultimate device performance. Two-dimensional growth is desirable because of the need for multilayered structures with flat interfaces and smooth surfaces. A notable exception is the fabrication of quantum dot devices, which requires three-dimensional or SK growth of the dots. Even here it is desirable for the other layers of the device to grow in a two-dimensional mode. In all cases of heteroepitaxy, it is important to be able to control the nucleation and growth mode.

3. Pseudomorphic and metamorphic growth

One of the main requirements for high quality heterostructure growth is the lattice constant of the growth material to be nearly the same as those of the substrate. In semiconductor alloys the lattice constant and band gap can be modified in a wide range. The lattice parameter difference may vary from nearly 0 to several per cent as in the cases of GaAs-AlAs and InAs-GaAs system, respectively. The growth of dilute nitride alloys is difficult because of the wide immiscibility range, a large difference in the lattice constant value and very small atom radius of N atoms. The growth of thick epitaxial layers creates many problems which absent in the quantum-well structures.

At the initial stage of the growth when the epitaxial layer is of different lattice constant than the substrate in-plane lattice parameter of the growth material will coherently strain in order to match the atomic spacing of the substrate. The elastic energy of deformation due to the misfit in lattice constant destroys the epilayer lattice. The substrate is sufficiently thick and it remains unstrained by the growth of the epitaxial layer. If the film is thin enough to remain coherent to the substrate, then in the plane parallel to the growth surface, the thin film will adopt the in-plane lattice constant of the substrate, i.e. $a_{\parallel} = a_o$, where $a_{\parallel}$ is the in-plane lattice constant of the layer and a_o is the lattice constant of the substrate. This is the case of pseudomorphic growth, and the epitaxial layer is pseudomorphic. If the lattice constant of the layer is larger than that of the substrate as in the case of InGaAs on GaAs, under the pseudomorphic condition growth the lattice of the layer will be elastically compressed in the two in-plane directions. The lattice constant of the layer in the growth direction perpendicular to the interface (the so-called out-of plane direction) will be strained according the Poison effect and will be larger than the unstrained value and the layer lattice will tense in the growth direction. Schematically this situation is illustrated in Figure 3.1.

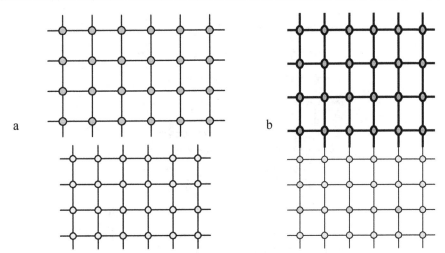

Fig. 3.1. Schematic presentation of atom arrangement for two materials with different cubic lattice constant: a) before growth; b) for pseudomorphic growth

In the case of the smaller lattice constant of the growth layer (GaAsN on GaAs for example), $a < a_0$ the layer will be elastically tensed in two in-plane directions and compressed in the growth directions (the out-of-plane lattice constant will be smaller than substrate lattice constant). Under pseudomorphic growth conditions the cubic lattice doesn't remain cubic: $a_{\parallel} = a_0 \neq a_{\perp}$. The out-of-plane lattice constant could be determined from the equation:

$$a_{\perp} = a[1 - D(a_{\parallel}/a - 1)] \tag{3.1}$$

Where:

$a_{\perp}$ - out-of-plane lattice constant of the layer

$a_{\parallel}$ - in-plane lattice constant of the layer

a - lattice constant of the unstrained cubic epitaxial layer

$D = 2C_{12}/C_{11}$, where C_{11} and C_{12} are elastic constants of the grown layer

Beyond a given critical thickness η_c when a critical misfit strain ε is exceeded, a transition from the elastically distorted to the plastically relaxed configuration occurs. In this case both mismatch component differ from zero: $a_{\parallel} \neq a_0 \neq a_{\perp}$. The lattice constant misfit is:

$$f = (a - a_0)/a_0$$

$$f_{\perp} = (a_{\perp} - a_0)/a_0 = (1 + D - DR)f \tag{3.2}$$

$$f_{\parallel} = (a_{\parallel} - a_0)/a_0 = Rf$$

R is a relaxation rate. For pseudomorphic growth $R=0$, and for full strain relaxation $R = 1$

If the epilayer is thicker than the critical thickness, there will be sufficient strain energy in the layer to create dislocations to relieve the excess strain. The layer has now returned to its unstrained or equilibrium lattice parameters in both the in-plane and out-of-plane directions and the film to be 100% relaxed. Figure 3.2 shows schematically how a misfit dislocation can relieve strain in the heteroepitaxial structure.

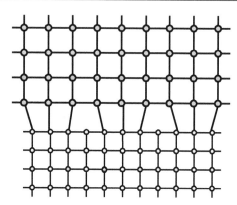

Fig. 3.2. Schematic presentation of the atom arrangement for metamorphic growth

In actual films, there is usually some amount of partial relaxation, although it can be very small in nearly coherent layers and nearly 100% in totally relaxed layers. For the partially relaxed layer, the in-plane lattice constant has not relaxed to its unstrained value. So some mismatch is accommodated by elastic strain, but a portion of the mismatch is accommodated by misfit dislocations (plastic strain).

There are two widely used models for calculations the critical thickness values: the Matthews-Blakeslee mechanical equilibrium model (Matthews.& Blakeslee, 1974) and the People-Bean energy equilibrium model (People & Bean, 1985). The People-Bean energy equilibrium model requires the total energy being at its minimum under critical thickness. According this model the elastic energy is equal to the dislocation energy at the critical thickness if the total elastic energy of the system with fully coherent interface is larger than the sum of the total system energy for the reduced misfit, due to the generation of dislocations, and the associated dislocation energy, and then begins the formation of interfacial dislocations.

Generally, the Matthews-Blakeslee model based on stemming from force balance, is the most often used to describe strain relaxation in thin films system. The equilibrium model of Matthews-Blakeslee assumes the presence of threading dislocations from the substrate. It gives mathematical relation for critical thickness by examining the forces originating from both the misfit strain F_ε and the tension of dislocation line F_L. The critical thickness h_c is defined as the thickness limit when the misfit strain force F_ε is equal to the dislocation tension force F_L(at h_c $F_\varepsilon = F_L$). For layers ticker than the critical thickness, the threading segment begins to glide and creates misfit dislocations at the interface to relieve the mismatch strain. The dislocations can easily move if dislocation lines and the Burgers vectors belong to the easy glide planes as {111} planes in face-centred cubic crystals.

In III-V semiconductors, the relaxation is known to occur by the formation of misfit dislocations and /or stacking faults. The usual misfit dislocations that are considered are located along the intersection of the glide plane and the interface plane. In zinc-blende crystal structures, on (100) oriented substrates the glide planes intersect the interface (110) which provides the corresponding line directions of misfit dislocations in such structures. The component of 60° dislocations perpendicular to the line directions contributes to strain relaxation. The 60° Burgers vector is b= ½ a_1 $\langle 110 \rangle$ and has a length along the interface perpendicular to the line $a / \sqrt{2}$.

Calculated values for critical thickness from People-Bean energy equilibrium and Matthews-Blakeslee force balance models are:

$$h_c = \frac{b(1-v)}{32\pi f^2(1+v)} \ln(h_c / b)$$ (3.3)

$$h_c = \frac{b}{4\pi f(1+v)}[\ln(h_c / b) + 1]$$ (3.4)

Where:

$v = C_{12}/(C_{12} + C_{11})$ is Poison's ratio,

f is a lattice mismatch, $b = a / \sqrt{2}$ is a magnitude of Burgers vector

The calculated values of People-Bean models are larger than that of the Matthews-Blakeslee model. The measurements of dislocation densities in many cases showed no evidence of misfit dislocations for layer considerable ticker than Matthews-Blakeslee limit and nearly close to the energy-equilibrium thickness limit. Layers with thicknesses above the People-Bean limit can be considered to be completely relaxed, whereas layers below Matthews-Blakeslee limit values fully strained. Layers with thicknesses between these limits are metastable. They could be free of dislocations after growth, but are susceptible to relaxation during later high-temperature processing.

For the semiconductor devices based on the thick metamorphic structure the influence of the misfit dislocations which are located at the interface on active region could be reduced by growing the additional barrier layers before active region growth. Threading dislocations, which propagate up through the structure, are the most trouble for electronic devices since they can create defect states such as nonradiative centres and destroy the device properties.

There are a variety of techniques used to reduce the density of threading dislocations in a material. For planar structure a thick buffer layer with lattice parameter equal to that of the active layers is usually used for reduction of threading dislocations. However, these structures always have high threading dislocation densities. In most thick nearly relaxed heteroepitaxial layers, it is found that the threading dislocation density greatly exceeds that of the substrate. Some authors (Sheldon et al.1988, Ayers et al. 1992) are noted for a number of heteroepitaxial material systems that this dislocation density decreases approximately with the inverse of the thickness. The dislocation density could be reduced by postgrowth annealing.

A linearly graded buffers and graded superlattice also are effectively used for restricting dislocations to the plane parallel to the growth surface, and thus support the formation of misfit dislocation and suppress threading dislocation penetration in the active region.

3.1 X-ray diffraction characterization

The X-ray diffraction (XRD) method is an accurate nondestructive method for characterization of epitaxial structures. X-ray scans may be used for determination the lattice parameter, composition, mismatch and thicknesses of semiconductor alloys.

In XRD experiment a set of crystal lattice planes (hkl) is selected by the incident conditions and the lattice spacing d_{hkl} is determined through the well-known Brag's law:

$$2d\sin\theta_B = n\lambda$$ (3.1.1)

where n is the order of reflection and θ_B is the Brag angle

The crystal surface is the entrance and exit reference plane for the X-ray beams in Bragg scattering geometry and the incident and diffracted beams make the same angle with the lattice planes. Two types of rocking curve scan are used: symmetric when the Bragg diffraction is from planes parallel to crystal surface and asymmetric when the diffraction lattice planes are at angle φ to the crystal surface (Fig. 3.3).

Fig. 3.3. Symmetric and asymmetric reflections from crystal surface

Let ω be the incidence angle with respect to the sample surface of a monochromatic X-ray beam. By rocking a crystal through a selected angular range, centered on the Bragg angle of a given set of lattice planes a diffraction intensity profile $I(\omega)$ is collected. For single layer heterostructure, the intensity profile will show two main peaks corresponding to the diffraction from the layer and substrate. The angular separation $\Delta\omega$ of the peaks account for the difference Δd_{hkl} between the layer and substrate lattice spacing. XRD do not directly provide the strain value on the crystal lattice. Te measurable quantities being the lattice mismatches $\Delta a_\perp/a_0$ and $\Delta a_{\parallel}/a_0$, i. e $f_\perp$ and $f_{\parallel}$. The relationship between lattice mismatch components and misfit f with respect to substrate is:

$$f = f_\perp\,(1\text{-}v)/(1+v) + 2\,v\,f_{\parallel}/\,(1+v) \tag{3.1.2}$$

where v is the Poisson ratio

This is the basic equation for the strain and composition characterization of heterostructures for cubic lattice materials. In the case of semiconductor alloys A_xB_{1-x} the composition x can be obtained if the relationship between composition and lattice constant is known. Poisson ratio is also composition depending and the use of Poisson ratio v is only valid for isotropic materials. For a cubic lattice, it can only be applied for high symmetric directions as (001), (011), (111), but Poison ratio may be different along different directions (v ≈ 1/3 for the most semiconductors alloys).

XRD can easily be employed to measure the lattice parameter with respect the substrate used as a reference. The strain and the composition of layer can be accurately determined if the dependence of the lattice parameter with the composition is known, the accuracy being mainly due to the precise knowledge of the lattice parameter –composition dependence.

In many cases a good approximation of a such dependence is given by Vegard law, which assumes that in the alloy A_xB_{1-x} the lattice of the alloy is proportional to the stoichiometric coefficient x:

$$a\,(x) = xa(A) + (1\text{-}\,x)\,a(B) \tag{3.1.3}$$

From this equation the stoichiometric coefficient x is obtained:

$$x = (a(x) - a(B))/ (a(A) - a(B)) \qquad (3.1.4)$$

If $a(B)$ is the substrate lattice parameter, the composition x can be calculated from the measurement misfit $f(x)$ value:

$$x = f(x)/f(AB) \qquad (3.1.5)$$

Where:

$f(x)$ is the measured misfit value with respect to $a(B)$ and

$f(AB)$ is the misfit between compound A and compound B, used as reference.

In the case of $GaAs_{1-x}N_x$ and $In_xGa_{1-x}As_{1-y}N_y$ dilute nitride alloys relationship between lattice parameters and composition assuming Vegard's law are the foolowing:

$$a_{GaAs_{1-x}N_x} = x\, a_{GaN} + (1-x)\, a_{GaAs} \qquad (3.1.6)$$

$$a_{\,In_xGa_{1-x}As_{1-y}N_y} = x\, y a_{InN} + (1-x)y\, a_{GaN} + x(1-y)\, a_{InAs} + (1-x)(1-y)\, a_{GaAs} \qquad (3.1.7)$$

The lattice parameter measurements method is one of the most accurate way to determine the composition, provided that the composition versus lattice parameter dependence is known. The comparison between composition values obtained from XRD and that, determined by other analytical techniques has allowed to measure the deviation from the linear Vegard's law in alloys.

Table 1. presents the values of elastic constants and lattice parameters for GaAs, InAs, GaN, InN binary compounds.

compound / Parameter	GaAs	InAs	GaN	InN
C_{11}, GPa	118.79	83.29	293	187
C_{12}, GPa	53.76	45.26	159	125
a_0, nm	0.5653	0.60584	0.4508	0.4979

Table 1. Elastic constants and lattice parameters for some III-V compounds

4. Low-temperature LPE growth

Low-temperature LPE is the most simple, low cost and safe method for high-quality III-V based heterostructure growth. It remains the important growth technique for a wide part of the new generations of optoelectronic devices, since the competing methods, MBE and MOCVD, are complicated and expensive although they offer a considerable degree of flexibility and growth controllability. The lowering the growth temperature for Al-Ga-As system provides the minimal growth rate values of 1–10 Å/s, and they are comparable with MBE and MOCVD growth values (Alferov et al, 1986). At the early stages of the process two-dimensional layer growth occurs, which ensures structure planarity and makes it possible to obtain multilayer quantum well (QW) structures (Andreev et al, 1996).

The results of study the crystallization process in the temperature range 650-400 ℃ demonstrate precise layer composition and thickness controllability for the low-temperature LPE growth. A necessary requirements for successful devices fabrication is the optimal doping of the structure layers at low temperatures. The experiments (Milanova and Khvostikov, 2000) on doping using different type dopants covered large range of carrier concentrations: from 10^{16} to 10^{19} cm^{-3} for n $Al_xGa_{1-x}As$ layers ($0 \leq x < 0.3$); from 5×10^{17} cm^{-3} to well above 10^{19} cm^{-3} for p $Al_xGa_{1-x}As$ ($0 \leq x < 0.3$); and from 10^{16} to 10^{18} cm^{-3} for n- and p $Al_xGa_{1-x}As$ ($0.5 < x < 0.9$) layers. High quality multilayer heterostructures containing layers as thin as 2-20 nm, as well as several microns thick, with a smooth surface and flat interfaces have been grown by low-temperature LPE. The lowest absolute threshold current of 1.3mA (300 K) was obtained for buried laser diodes with a stripe width of ~ 1μm and cavity length of 125μm (Alferov et al, 1990).

High-efficiency solar cells for unconcentrated (Milanova et al, 1999) and concentrated solar cells (Andreev et al, 1999) have been fabricated by low-temperature LPE. The record conversion efficiency under ultra-high (>1000) concentration ratio solar radiation heve been achieved for GaAs single-junction solar cells based on multilayer AlGaAs/GaAs heterostructures (Algora et al, 2001).

The success of the LPE method is strongly depend on the graphite boat design used for epitaxy growth. The most widely used for LPE growth is a slide boat method. The conventional simple slide boat consists of a boat body in which are formed containers for liquid phase and a slider with one or more sits for the substrate (Fig. 4.1.). The slider moves the substrates under and out of the growth melt. This boat design has some disadvantages: the melt thicknesses is several millimeters and during growth from such semi-limited liquid-phase a portion of dissolved materials can not reach the substrate surface and forms stable seeds at a distance of 1 mm and more from the growth surface which deteriorate the planarity of the grown layer.

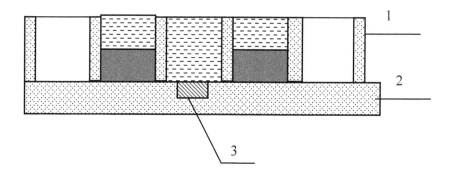

Fig. 4.1. Conventional slide boat for LPE growth: 1, body boat; 2, slider, 3, substrate.

Another drawback is the arising the defects on the layer surface due to the mechanical damage during its transfer from one melt to another. Also always on the surface of the melt present oxides films and it is difficult to completely removed these films even by long high-temperature baking. This is a critical problem for wetting of the substrate surface, especially for epitaxial process in Al-Ga-As system. A piston growth technique has been developed for LPE growth of AlGaAs heterostructures by Alferov at al (Alferov et al, 1975).

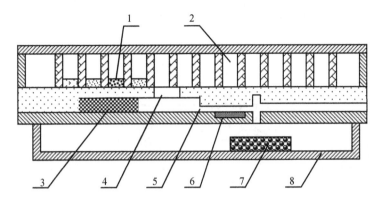

Fig. 4.2. Piston boat for growth of multilayer AlGaAs/GaAs heterostructures: 1, growth solution; 2, container for solution; 3, piston; 4, opening; 5, narrow slit; 6, substrate; 7, used solution; 8, container for used solutions.

The substrate surface in this boat after the first wetting is always covered by a melt and this solves difficulties of wetting during the growth of AlGaAs heterostructures in the range 600-400 °C. The piston boat design is shown in Figure 4.2. In this boat the melts of different compositions are placed in containers which can move along the boat body. The liquid phase falls down into the piston chamber and squeezes throw narrow slit into the substrate which allows mechanical cleaning of oxides films from liquid phase and insures a good wetting. The crystallization is carried out from the melt 0.5-1 mm thick. After the growth of the layer liquid phase is removed from the substrate by squeezing of the next melt. The last liquid phase is swept from the surface by shifting the substrate holder out side the growth chamber.

The liquid phase can not remove completely from the surface structure and cause a poor morphology of the last grown layer. The excess melt could be remove from the substrate by using additional wash melt, which may either has a poor adhesion to the substrate or may be relatively easy remove with post-growth cleaning and etching in selective etchants. Authors (Mishurnyi at al, 2002) suggest an original method to complete remove the liquid solution after epitaxy. The remained liquid phase is pulled up into the space between the substrate and vertical plates made of the same materials as the substrate assembled very closely to the substrate surface. This method is very useful for growth of multilayer heterostructures not containing Al in modified slide boat because prevent mixing of any liquids remaining. For the most multicomponent alloys such as InGaAsP, InGaAsSb, InPAsSb etc., lattice constant is very sensitive to composition variation and the piston boat is not suitable for their growth because of mixing of two deferent solutions. Slide boats with different design are used for fabrication of complicated multilayer heterostructure on the base of these multicomponent alloys. In order to improve the control of layer thicknesses and uniformity it is necessary the growth to be carried out using a finite melt. In this boat the liquid phase after saturations is transferred into the additional containers or growth chamber with finite space for the liquid phase. Figure 4.3 shows a schematic slide boat for epitaxy growth from finite melt. A critical requirement for the most multicomponent alloys, instead of AlGaAs, AlGaP, is precise determination of the growth temperature. The temperature at the interface between the liquid phase and substrate can not be measured and common it is determined by measurements of the source component solubility (Mishurnyi et al, 1999) .

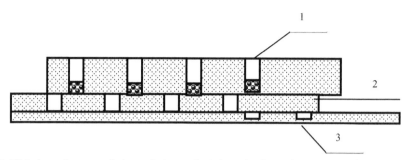

Fig. 4.3. Slide boat for growth from finite melt: 1, boat body with container for melts; 2, slider with container for finite melts; 3, slider for the substrates.

Slide boats with different design modification are used for growth of variety structures in different multicomponent system. A boat made of two different materials, sapphire (for body) and graphite (for slider), is suggested by Reynolds and Tamargo (Reynolds and Tamargo, 1984). This design reduces temperature variations around the perimeter of the substrate which contribute to unwanted 'edge' growth effects. Slide boats with narrowed melt contact for epitaxy of extremely thin epilayers have been used to grow active layer in single-quantum well lasers by (Alferov et al, 1985) and later by (Kuphal, 1991). Also a modified slide boat can be used for multilayer periodic structures growth (Arsent'ev et al, 1988). The use of two growth chambers with narrow slits makes it possible to produce such structures by means of repeated reciprocating movements of the slider with the substrate situated underneath these slits. Another variant of an LPE boat (Mishurnyi et la, 1997), which is a combination of the 'sliding' and 'piston' designs has been used successfully to grow InGaAsSb, AlGaAsSb and various multilayer structures on the basis of these materials.

5. Low-temperature LPE growth and characterization of dilute nitride GaAsN and InGaAsN thick layers

Dilute nitride III-V-N alloys with nitrogen content in the range of few percent, such as GaAsN and InGaAsN, are of considerable interest for application in multijunction solar cells.

The incorporation of nitrogen into group V sublattice causes profound effect on the band gap and properties of the dilute nitride material strongly differ from those of the conventional III-V alloys. While in conventional alloys a smaller lattice constant increases the band gap, the mixing of GaAs with few molar percent of GaN leads to giant reduction of its band gap due to the smaller covalent radius and large electronegativity of N atoms. The large changes in the electronic structure in dilute III–V nitrides could be explained by the band anticrossing model (BAC). The interaction between the localized levels introduced by a highly electronegative impurity, such as N in GaN_xAs_{1-x}, and the delocalized states of the host semiconductor causes a restructuring of the conduction band into E+ and E− subbands, which in this case effectively lowers the conduction band edge of the alloy.

Figure 5.1. shows the relationship between the lattice constant and band-gap energy in some III-V semiconductor alloys. In the case of InGaNAs adding In to GaAs increases the lattice constant, while adding N to GaAs decreases the lattice constant. In the same time the incorporation of In and N in GaAs leads to reduction of the band gap energy in the new alloy. Consequently, by adjusting the contents of In and N in quaternary InGaNAs alloys can be grown lattice-matched to GaAs layers because In and N have opposing strain effects on the lattice and make it possible to engineer a strain-free band gap layers suitable for different applications.

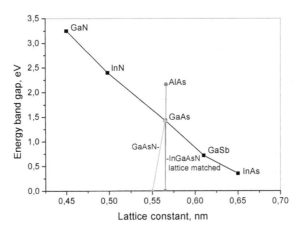

Fig. 5.1. Relationship between lattice constant and bad gap energy for some III-V semiconductor alloys

Recently a development of the spectral splitting concentrator photovoltaic system based on a Fresnel lens and diachronic filters has a great promise to reach super high conversion efficiencies (Khvostiokov et al. 2010). Module efficiency nearly 50% is expected for the system with three single-junction solar cells connected in series with band gap of 1.88-1.42-1.0 eV. The development of three optimized AlGaAs, GaAs and InGaAsN based cells is the best combination for application in such system if PV quality of the quaternary InGaAsN could be reached by LPE growth.

In this paper low-temperature LPE is proposed as a new growth method for dilute nitride materials. Because of its simplicity and low cost many experiments on GaInAsN and GaAsN growth under different condition and with different doping impurities could be made using LPE. The systematic study of their structural, optical and electrical properties by various methods make it possible to find optimized growth conditions for InGaAsN quaternary compounds lattice matched to GaAs substrate.

5.1 Growth and characterization of GaAsN layers

GaAsN compounds were grown by the horizontal graphite slide boat technique for LPE on (100) semi-insulating or n-type GaAs substrates. A flux of Pd-membrane purified hydrogen at atmospheric pressure was used for experiments. No special baking of the system was done before epitaxy. Starting materials for the solutions consisted of 99.9999 % pure Ga, polycrystalline GaAs and GaN. The charged boat was heated at 750°C for 1 h in a purified H_2 gas flow in order to dissolve the source materials and decrease the contaminants in the melt. Epitaxial GaAsN layers 0.8-1.5 thick were grown from different initial temperatures varied in the range 560-650 °C at a cooling rate of 0.6 °C/min.

5.1.1 Structural characterization

XRD and SIMS techniques are used to determine N concentration in grown samples. While SIMS measures the total nitrogen content in the layer, XRD determines the change in the lattice constant due to the substitution of nitrogen atoms on As-sublattice sites.

The N composition from XRD results could be estimated assuming Vegard's law. In many cases the Vegard's law is a good approximation for the lattice parameter dependence on the composition. The deviation from Vegard's law dependences on many parameters, for instance, the difference in the atom bond length, different atom electronegativity and elastic constants of the components in the alloy. For the ideal case N incorporates predominantly as substitutional N_{As} atoms in As- sublattice substituting As atoms. However, it is known that there are some other N configurations: N-As split interstitial; N-N split interstitial; and isolated N interstitial. Figure 5.2 presents the main configurations of N in GaAsN as substitutional atom N_{As} and as As-N and N-N split interstitials, respectively.

The influence of these N-related complexes on the lattice constant can be calculated on the base of the theoretical model of Chen (Chen et al. 1996) for analyzing the correlation between lattice parameters and point defects in semiconductors. According this model the lattice strain caused by the substitutional N_{As} is given by the following relation:

$$\frac{\Delta a}{a} = \mu \frac{x(r_N - r_{As})}{2(r_{Ga} + r_{As})} \tag{5.1.1}$$

where: $r_N, r_{Ga,} r_{As}$ the covalent radii;
$\mu = (1 + v)/(1 - v)$, and v is the Poissn ratio
The lattice strained caused by split interstitial is:

$$\frac{\Delta a}{a} = \mu \frac{x(d_b - r_{Ga} - r_{As})}{2(r_{Ga} + r_{As})} \tag{5.1.2}$$

Where d_b is the distance of the N-As complex from its nearest neighbours:

$$d_b = \frac{\sqrt[3]{3}}{3} r_{si} + \sqrt{(r_{si} + r_{Ga})^2 + \frac{2}{3} r_{si}^2} \tag{5.1.3}$$

where $r_{si} = (r_N + r_{As})/2$ is an effective bond radius.

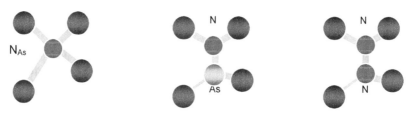

Fig. 5.2. The main configurations of nitrogen atoms in GaAsN
● N-atom,   ● As-atom ● Ga-atom

The effect of N-N interstitial is very small and can be neglected. Also the formation of an isolated N interstitial is unlikely due to a high formation energy (Li et al. 2001) and their concentrations in GaAsN is very small. While the substitutional N_{As} atoms compress the lattice constant , the N-As complexes expand the lattice constant of GaAsN in the growth direction, as shown in Figure 5.3. So, XRD results may underestimated the N composition due to the N-As and N-N split interstitials.

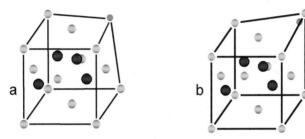

Fig. 5.3. Incorporation of N-atom in As-sublattice: a) as substitutional atom N_{As}; b) as As-N split interstitial

XRD rocking curves are recorded in the symmetrical (004) reflection. Fig. 5.4 shows the experimental XRD rocking curves of two GaAsN samples, 1.2 μm thick, with N composition of 0.3% and 0.62% and may consider that they are fully relaxed. The N content determines the line shape of the main peak of the spectra: it manifests itself as a broad shoulder evolving into a weak separate peak shifted away to the right from the (004) GaAs substrate reflection. Our data show that N compositions measured by the two methods, SIMS and XRD, agree well and XRD measurements by using Vegard's law could be used to determine the lattice constant of GaAsN layers containing low N concentrations. These results are in a good agreement with the calculations from the theoretical model and the experimental results for small N concentration in the GaAsN reported in the literature. The deviation from Vegard's law has been observed for nitrogen concentration levels above 2.9 mol % GaN in the layer (Spruytte at all., 2001; Li et al. 2001).

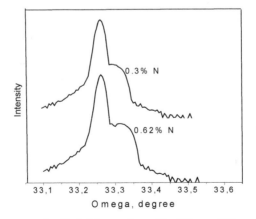

Fig. 5.4. XRD rocking curves for GaAsN samples with differenrt N content.

Unlike XRD used for assessing the incorporation of nitrogen in GaAs$_{1-x}$N$_x$ alloys grown by LPE the nitrogen bonding configurations and local atomic structures have been studied using x-ray photoelectron spectroscopy (XPS) and Fourier transform infrared (FTIR) spectroscopy. The XPS spectra have been measured over a range of binding energies from 1 to 550 eV. The X-ray photoelectron spectra of N 1s photoelectron and Ga LMM Auger lines recorded from the as grown GaAs$_{1-x}$N$_x$ samples prepared in different temperature ranges are shown in Fig 5.5. It is clearly seen the Ga Auger peak around 391 eV and the N 1s level photoemission peak of the samples. The variation of the intensity of the N 1s peak with respect to the Ga LMM peaks reflects is due to the different nitrogen content of the samples. Sample grown from higher initial epitaxy temperature of 650 °C contains 0.2% N and exhibits a N 1s peak with lower intensity and lower binding energy in comparison with the N 1s peak intensity of the sample grown in the lower temperature range (600-570 °C) with 0.5% N content. It has been established that lower epitaxy temperatures favours nitrogen incorporation in the layers. The N 1s spectra of the samples indicate that nitrogen atoms exist in a single-bonded configuration, the Ga-N bond, and interstitial nitrogen complexes is not observed, in contrast to data of high nitrogen content GaAsN samples where the additional nitrogen complex associated peak is recorded (Spruytte at all., 2001).

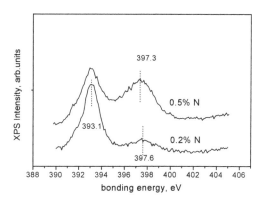

Fig. 5.5. XPS spectra of two GaAsN samples with different N content.

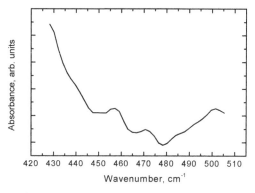

Fig. 5.6. FTIR spectrum of as grown GaAsN sample.

FTIR absorption spectra of an as grown GaAs$_{1-x}$N$_x$ layer on a n-GaAs substrate is plotted in Fig. 5.6. A peak at 472.6 cm^{-1}, attributed to a local vibrational mode of nitrogen at arsenic site in GaAs is clearly seen.

5.1.2 Electrical characterization

Electrical parameters of undoped GaAs and GaAsN layers with different nitrogen content grown on seminsulating (001) GaAs substrates are measured in the temperature range 80 – 300 K using van der Pauw geometry.

Figure 5.7. shows the temperature dependence of the Hall-concentration n_H on reciprocal temperature for two layers GaAsN with nitrogen concentration of 0.2% and 0.5%, respectively in comparison with undoped GaAs. It is seen that all samples are of n-type and for layers containing nitrogen electron concentration increases about one order of magnitude. This could be explained by the assumption that nitrogen behaves mainly as an isoelectronic donor, which arises from the local heterojunction scheme GaAs-GaN according to Belliache (Bellaiche et al., 1997). The results shown in figure indicate that the free carrier concentration increases strongly with the N concentration. The increase in n_H has also been observed in GaN$_x$As$_{1-x}$ doped with S (Yu et al., 2000a) and in Ga$_{1-3x}$ In$_{3x}$N$_x$As$_{1-x}$ alloys doped with Se (Skierbiszewski at al., 2000). This large increase of the free electron concentration can be quantitatively explained by a combination of the band anticrossing model (Shan et al, 1999) and the amphoteric defect model (Walukiewicz, 1989). The later suggests that the maximum free carrier concentration in a semiconductor is determined by the Fermi energy with respect to the Fermi-level stabilization energy E$_{FS}$ which is a constant for III-V semiconductors. Since the position of the valence band in GaAsN is independent of N concentration, the giant downward shift of the conduction band edge toward E$_{FS}$ and the enhancement of the density of states effective mass in GaAsN lead to much larger concentration of uncompensated, electrically active donors for the same location of the Fermi energy relative to E$_{FS}$. In order to explain the large enhancement of the doping limits in dilute nitride alloys both the effects of band gap reduction and the increase in the effective mass have to be taken into account (Yu et al., 2000 b; Skierbiszewski at al., 2000).

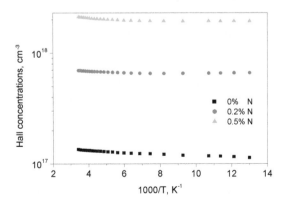

Fig. 5.7. Free carrier concentration as a function of inverse temperature for as grown GaAs, and two GaAsN layers with different N content

Figure 5.8. presents the temperature dependencies of the Hall-mobility for the same samples. The mobility of the dilute GaAsN samples is considerably lower due to space charge scattering contributions induced by N-related defects added to well-known scattering mechanisms such as phonon and ionized impurity scattering. The mobility maximums of both curves are almost at the same temperature with a relatively small difference of about 20K, which is an indication for scattering specificity. It is seen a well expressed low-temperature mobility decrease which could be explained by the temperature dependence of the GaAs conduction band edge energy, which is closer to the N defect levels at lower temperatures, increasing the scattering cross-section.

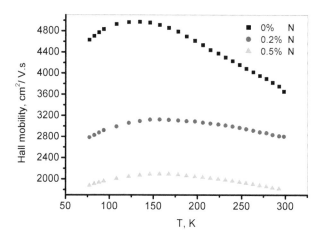

Fig. 5.8. Temperature dependence of Hall electron mobility for GaAs (full squares), and two GaAsN with: 0.2%N (full circles); 0.5%N (full triangles)

The mobility values of the dilute GaAsN samples is lower than those of the undoped GaAs layer but considerable higher than mobility values obtained in n-type GaAsN films with similar free electron concentration grown by MOCVD and MBE.

5.2 Growth and characterization of InGaAsN layers
Dilute InGaAsN layers have been prepared using the same technique as for GaAsN growth. A series of nearly-lattice matched InGaAsN epilayers 1.3-1.5 μm thick have been grown from In-rich solution containing 1.5 at.% polycrystalline GaN as a nitrogen source in the temperature range 615 – 580 °C at a cooling rate 0.6 °C/min.

5.2.1 Structural characterization
Typical XRD rocking curves for grown layers are plotted in the Fig. 5.9.
Two prominent peaks associated with the GaAs substrate and the quaternary InGaAsN layer are observed. The lattice mismatch $\Delta a/a_0$ determined from the XRD spectrum is ~ 0.1%. The In-concentration of the layers measured separately by X-ray microanalyses is 6.4%. Using Vegard's law the N- content in InGaAsN layers is determined to be 2.8%

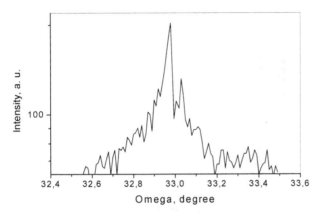

Fig. 5.9. XRD rocking curves of InGaAsN sample

The local structure of the InGaAsN is defined by Raman spectrum. In the Raman spectrum, presented in fig. 5.10. does not observed N-induced local mode LO2 , assigned to the vibration of isolated nitrogen atom bonded to four Ga neighbors ($N_{As}Ga_4$). Instead of this two LVM peaks at 454 and 490 cm^{-1} originated from In-N bonds in a local Ga_3In_1-N or Ga_2In_2-N configurations are appeared. Similar LVM peaks have been reported in the literature for as grown InGaAsN layers by MBE (Mintairov et al., 2001, Hashimoto et al., 2003) and for some MBE and MOCVD samples after annealing (Pavelescu et al.,2005; Kurtz et al. 2001). The experimentally observed local modes could be explained by theoretical analyses of the microscopic lattice structures related to the incorporation of N in InGaAsN alloys. The Monte Carlo simulation (Kim & Zunger, 2001) reveal that in InGaAsN quaternary alloys the "small atom–large atom" bond configuration i.e. "large cation-small anion" In-N + "small cation-large anion" Ga-As is preffered for better lattice-matched of the alloy to GaAs substrate, because introduces less strain. On the other hand, the cohesive energies of GaN is larger than that of InN, so the highly strained Ga-N + In-As configuration is preferred in terms of bond energy. In LPE growth under near to equilibrium conditions In-N bonds are more favorable since they reduce the sum of local strain plus chemical bond energies. The introduction of In changes N environment by formation short-range-ordered nitrogen centered N-In_nGa_{4-n} ($0 \leq n \leq 4$) clusters in InGaAsN alloy. In Ga-rich InGaAsN quaternary the most probably realized are the nearest –neighbor pair defects N_{As}-In_{Ga} in which one of the Ga atom in the neighborhood of N is replaced by a large size heavier In_{Ga} ($N_{As}In_{Ga}Ga_3$) and also a formation of a second nearest-neighbor complex $N_{As}In_{Ga}(2)Ga_2$ where two of four Ga atoms is replaced by two large-site and heavier In_{Ga} in the vicinity of N_{As}. The calculations using Green's function technique (Talwar, 2007) relieve the splitting of a triple degenerate N_{As} near to 471 cm^{-1} into a non-degenerate LVM ~ 462 cm^{-1} and a double degenerate LVM at 490cm^{-1} for the nearest –neighbor complex and three bands near to 481, 457, and 429 for second nearest-neighbor complex . The surface roughness of the samples has been examined by atomic force microscopy (AFM). A three-dimensional AFM image of an as grown 1.3 μm-thick InGaAsN layer is presented in Fig. 5.11. The measured root-mean-square (RMS) roughness on 1-micron area is 0.42 nm.

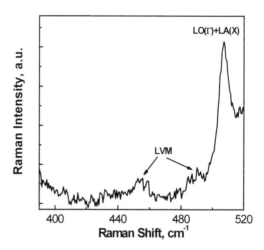

Fig. 5.10. Raman spectrum of as grown InGaAsN layer.

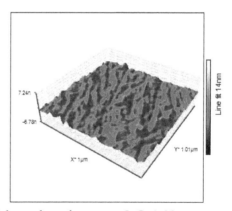

Fig. 5.11. AFM image of the surface of as grown InGaAsN.

5.2.2 Electrical characterization

In the Fig. 5.12 are plotted the temperature dependence of Hall concentrations n_H for lattice matched InGaAsN in comparison with a metamorphic InGaAs layer. For undoped InGaAs, n_H decreases linearity in the explored temperature range, 80 to 300K, typical for slightly degenerate III-V semiconductors. However, for N-containing films, two distinct temperature regimes with different temperature dependence of n_H are observed. The Hall electron concentration decreases as the temperature decreases down to about 200 K, indicating the presence of thermally activated deep donor levels within the dilute nitride bandgap. The saturation of n_H at low temperature (T < 200K) is attributed to fully ionized shallow donors. This behavior could be explained by the presence of two donor levels in the InGaAsN bandgap, one being a shallow N isoelectronic donor and the second a thermally activated deeper donor, presumably N-related deep-level defects typically associated with different N-N pair and N-cluster states (Zhang & Wei, 2001).

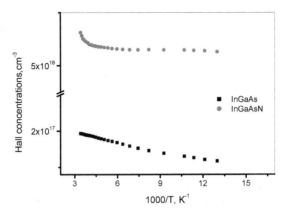

Fig. 5.12. Temperature dependence of free carrier concentrations for undoped InGaAs and lattice matched InGaAsN samples

The temperature dependence of Hall mobility for undoped InGaAs and InGaAsN layers grown from In-rich solution is similar to those for the GaAsN layers grown from Ga-rich solution as it is shown in Figures 5. 13 and 5. 14.

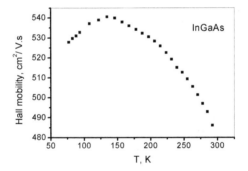

Fig. 5.13. Hall mobility as a function of temperature for undoped InGaAs sample

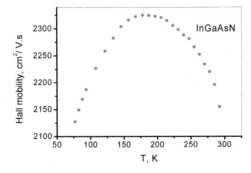

Fig. 5.14. Hall mobility as a function of temperature for lattice matched InGaAsN sample

The mobility of the metamorphic InGaAs structure is low, down to 500 cm^2/V.s, since it possibly contains threading dislocations of high density and the latter causes relatively poor material quality. High values over 2000 cm^2/ V.s for Hall mobility exhibits the lattice matched to GaAs substrate InGaAsN sample. These values are about the theoretical limit predicted by Fahy and O'Reilly (Fahy and O'Reilly, 2004) and among the highest reported for lattice matched thick InGaAsN layers.

6. Conclusion

Dilute nitride GaAsN and InGaAsN epitaxial layers have been prepared by low-temperature LPE using polycrystalline GaN as a source for nitrogen. The GaAsN layers, 0.8-1.5 μm thick, with 0.15-0.6 at. % N content in the solid have been grown from different initial epitaxy temperature varied in the range 650-550 °C. The lowering the epitaxy temperatures favors nitrogen incorporation in the layers. The Hall measurements reveal sharply increase of free carrier concentrations about one order of magnitude and decrease of Hall mobility for GaAsN samples in comparison with undoped GaAs.

Lattice-matched conditions for coherent growth of InGaAsN layers on GaAs have been found. The results suggest preferential In-N bond formation for high quality growth of these alloys. Temperature dependent electronic transport measurements show a thermally activated increase in the free carrier concentration at measurement temperatures higher than 200 K, suggesting the presence of carrier trapping levels below the GaAsN conduction band edge. Nearly lattice matched to GaAs substrate thick $In_xGa_{1-x}As_{1-y}N_y$ (x~ 6.4%, y~2.8%) layers exhibit high values over 2000 cm^2/ V.s for Hall electron mobility.

Further study is necessary in order to determine the potential of melt-grown quaternary InGaAsN alloys for solar cell application. This will be attained by: study the influence of growth conditions on the material quality in wide temperature range 450-600 °C and differentiation between intrinsic and extrinsic limitations for device performance; finding optimized growth conditions for InGaAsN lattice matched to GaAs with improved material quality ; extending the long-wavelength limit of GaAsN-based materials by the lowering the band gap energy of dilute nitride structure that can be lattice matched grown on GaAs.

The finale goal is a development of single-junction solar cells with high photovoltaic parameters.

7. References

Alferov Zh.I., Andreev V.M., Konnikov S.G., Larionov V.R. and Shelovanova G.N. (1975). Investigation of a new LPE method of obtaining Al-Ga-As heterostructures, *Kristall Technik*, Vol.10, No 2, pp 103-110.

Alferov Zh.I., Garbuzov D.Z., Arsent'ev I.N., Ber B.Ya., Vavilova L.S., Krasovskii V.V. and Chydinov A.V.(1985) Auger profiles of the composition and luminescence studies of liquid-phase grown InGaAsP

Alferov Zh.I., Andreev V.M., Vodnev A.A., Konnikov S.G., Larionov V.R., Pogrebitskii K.Yu. Rymiantsev V.D. and Khvostikov V.P.(1986). AlGaAs heterostructures with quantum-well layers, fabricated by low-temperature liquid-phase epitaxy, *Sov. Tech. Phys. Lett.*, 12(9), 450–451.

Alferov Zh.I., Andreev V.M., A.Z. Mereutse, Syrbu A.V., Suruchanu G.I. and Yakovlev V.P.(1990). Super low-threshold (Ith = 1.3 mA, T= 300 K) quantum-well AlGaAs lasers with uncoated mirrors prepared by liquid-phase epitaxy, *Sov. Tech. Phys. Lett.*, Vol.16, No 5, pp 339–340.

Algora C., Oritz E., Rey-Stolle I, Diaz V., Pena R., Andreev V. M., Khvostikov V. P., Rumyantsev V. D.(2001). A GaAs solar cell with an efficiency of 26.2% at 1000 suns and 25.0% at 2000 suns, *IEEE Trans. on Elec. Dev.*, Vol. 48, No 5, pp 840-844. ISSN: 0018-9383.

Andreev V. M., Kazantsev A. B., Khvostikov V. P., Paleeva E. V., Rumyantsev V. D. and Sorokina S. V. (1996). Quantum-well AlGaAs heterostructures grown by low-temperature liquid-phase epitaxy, *Materials Chemistry and Physics*, Vol. 45, No 2, p p.130-135.

Andreev V. M., Khvostikov V. P., Larionov V. R., Rumyantsev V. D., Paleeva E. V., Shvarts M. Z.(1999). High-efficiency AlGaAs/GaAs concentrator (2500 suns) solar cells *Fiz. Tekh. Poluprovodn.* Vol. 33, N0 9, pp 1070-1072.

Arsent'ev I.N., Bert N.A., Vasil'ev A.V., Garbuzov D.Z., Zhuravkevich E.V., Konnikov S.G., Kosogov A.O., Kochergin A.V., Faleev N.N. and Flaks L.I.(1988). Periodic multilayer In-Ga-As-P structures fabricated by liquid-phase epitaxy, *Sov. Tech. Phys. Lett.*, Vol 14, No 4, pp 264–266.

Ayers J.E., Schowalter L.J., and S.K. Ghandhi (1992). Post-growth thermal annealing of GaAs on Si(001) grown by organometallic vapor phase epitaxy, *J. Cryst. Growth*, Vol. 125, No1-2, pp. 329-335.

Bauer, E . (1958). Phaenomenologische Theorie der Kristallabscheidung an Oberflaechen, *Zeitschrift für Kristallographie*, Vol. 110, pp 372-394.

Bellaiche L., S. H. Wei, Zunger A., (1997) Band gaps of GaPN and GaAsN alloys, *Appl. Phys. Lett.*, Vol. 70, No 26, pp.3558-3560.

Chen N. F., Wang Y., He H, and Lin L. (1996). Effects of point defects on lattice parameters of semiconductors, *Physical Review B*, Vol. 54, No 12, (September 1996), pp 8516-8521.

Dhar S., Halder N., Mondal A., Bansal B., Arora B. M. (2005). Detailed studies on the origin of nitrogen-related electron traps in dilute GaAsN layers grown by liquid phase epitaxy, *Semiconductor Science and Technology*, Vol. 20, No. 12. pp. 1168-1172.

Fahy S., O'Reilly E. P. (2004), Theory of electron mobility in dilute nitride semiconductors, *Physica E: Low-dimensional systems and nanostructures*, Vol. 21, No 1-2, pp. 881-885.

Frank F.C. and Van der Merwe J.H. (1949). One-dimensional dislocations. II. Misfitting monolayers and oriented overgrowth, *Proc. R. Soc. London A*, Vol. 198, No 1053, (August 15, 1949), pp 216 -225.

Grabow M.H. and Gilmer,G.H. (1988). Thin film growth modes, wetting and cluster nucleation, *Surf. Science*, Vol. 194, No 3, pp 333-346

Guter W., Schöne J., Philipps S., Steiner M., Siefer G., Wekkeli A., Welser E, Oliva E, Bett A. and Dimroth F.(2009). Current-matched triple-junction solar cell reaching 41.1% conversion efficiency under concentrated sunlight, *Appl. Phys. Lett.* Vol. 94, No 2, 023504.

Johnston S. W., Kurtz S. R., Friedman D. J., Ptak A. J., Ahrenkiel R. K., and Crandall R. S. (2005). Observed trapping of minority-carrier electrons in *p*-type GaAsN during deep-level transient spectroscopy measurement, Appl. Phys. Lett. Vol 86, No 7, 072109.

Hashimoto A., Kitano T., Nguyen A.K., Masuda A., Yamamoto A., Tanaka S., Takahashi M., Moto A., Tanabe T., Takagishi S., Raman characterization of lattice-matched

GaInAsN layers grown on GaAs (001) substrates, *Solar Energy Materials & Solar Cells* 2003, Vol. 75, pp. 313–317.

Hiramatsu K., Itoh S., Amano H., Akasaki I., Kuwano N., Shiraishi T. and Oki K.(1991). Growth mechanism of GaN grown on sapphire with A1N buffer layer by MOVPE, *J. Cryst. Growth*, Vol.115, No 1-4 , (Desember,1991), pp 628-633.

Khan A, Kurtz S. R., Prasad S., Johnston S. W., and Gou J. (2007). Correlation of nitrogen related traps in InGaAsN with solar cell properties, *Appl. Phys. Lett.* Vol. 90, No 24, 243509.

Khvostikov V.P., Sorokina S.V., Potapovich N.S., Vasil'ev V.I., Vlasov A.S., Shvarts M.Z., Timoshina N.Kh., Andreev V.M., (2010). Single-junction solar cells for spectrum splitting PV system, *Proceedings of the 25th European PV Solar Energy Conference and Exhibition*, Valencia, Spain6-10 September 2010.

Kim K. and Zunger A., Spatial Correlations in GaInAsN Alloys and their effects on Band-Gap Enhancement and Electron Localization, *Phys. Rev. Lett.*, Vol. 86, No 12, pp. 2609-2612.

King R.(2008). Multijunction cells: record breakers, *NaturePhotonics*, Vol. 2, pp. 284– 285.

Krispin P., Spruytte S. G., Harris J. S., and Ploog K. H. (2002) Electron traps in Ga(As,N) layers grown by molecular-beam epitaxy, *Appl. Phys. Lett.* Vol. 80, No 12, pp 2120-2122.

Kuphal E. (1991). Liquid Phase Epitaxy, *Appl. Phys. A* 52, pp 380–409.

Kurtz S. R, Allerman A. A., Jones E. D., Gee J. M., Banas J. J., and Hammons B. E. (1999) InGaAsN solar cells with 1.0 ev band gap, lattice matched to GaAs, *Applied Physics Letters*, Vol. 74, No 5, pp. 729–731.

Kurtz S. R., Allerman A. A., Seager C. H., Sieg R. M., and Jones E. D. (2000), Minority carrier diffusion, defects, and localization in InGaAsN, with 2% nitrogen, *Applied Physis Letters*, Vol. 77, No 3, pp. 400–402.

Kurtz S., Webb J., Gedvilas L., Friedman D., Geisz J., Olson J., King R., Joslin D., & Karam N., *Appl. Phys. Lett.*, 2001, Vol. 78, No 6, pp. 748 -750.

Kurtz S. R., Klem J. F., Allerman A. A., Sieg R. M., Seager C. H., & Jones E. D. (2002) Minority carrier diffusion and defects in InGaAsN grown by molecular beam epitaxy, *Appl. Phys. Lett.*, Vol. 80, No 8, pp. 1379-1381.

Li W., Pessa M., & Likonen J. (2001) Lattice parameter in GaNAs epilayers on GaAs: Deviation from Vegard's law, *Appl. Phys. Lett.*, Vol. 78, No 19, pp. 2864-2866.

Mattes B.L. and Route R.K.(1974). LPE growth of GaAs: Formation of nuclei and surface terraces , *J. Cryst. Growth*, Vol. 27, (December 1974), pp 133-141.

Matthews J. W.& Blakeslee A. E. (1974). Defects in epitaxial multilayers: I. Misfit dislocations, *J. Cryst. Growth*, Vol. 27, (December 1974) pp. 118-125.

Milanova M., Mintairov A., Rumyantsev V., Smekalin K.(1999). Spectral characteristics of GaAs solar cells grown by LPE, *J. of Electronic Mat.*, Vol 28, No 1, pp 35-37.

Milanova M. and Khvostikov V.(2000). Growth and doping of GaAs and AlGaAs layers by lowtemperature liquid-phase epitaxy, *J. Crystal Growth*, Vol. 219, No 10, pp 193–198.

Milanova M., Kakanakov R., Koleva G., Arnaudov B., Evtimova S., Vitanov P., Goranova E.,. Bakardjieva Vl, Alexieva Z.(2009) Incorporation of nitrogen in melt grown GaAs , *J. Optoelectronics and Advanced Materials*, Vol. 11, No. 10, pp. 1471-1474.

Mintairov A. M., Blagnov P. A.,. Merzj J. L, Ustinov V. M. & Vlasov A. S. (2001). Vibrational study of nitrogen incorporation in InGaAsN alloys, *9th Int. Symp. "Nanostructures: Physics and Technology"* NC.13p, St Petersburg, Russia, June 18-22, 2001.

Mishurnyi V.A., de AndaF., Gorbatchev A.Yu., Vasil'ev V.I., Faleev N.N.(1997). InGaAsSb growth from Sb-rich solutions, *J. Crystal Growth*, Vol. 180, pp 34–39.

Mishurnyi V.A., de Anda F., Hernandes del Castillo I.C. and Gorbatchev A.Yu.(1999). Temperature determination by solubility measurements and a study of evaporation of volatile components in LPE, *Thin Solid Films*,Vol. 340, pp 24–27.

Mishurnyi V.A., de Anda F. and Gorbatchev A.Yu.(2002). Some problems on the phase diagrams research in LPE, *Current Top. Crystal Growth Res.*, Vol. 6, pp 115–125.

Nabetani Y., Ishikawa T., Noda S. and Sasaki A.(1994). Initial growth stage and optical properties of a three-dimensional InAs structure on GaAs, *Journal of Applied Physics*, (July, 1994) Vol. 76, No 1, pp 347-351.

Nishimura K., Lee H.-S. , Suzuki H., Ohshita Y., & Yamaguchi M. (2007) Chemical Beam Epitaxy of GaAsN Thin Films with Monomethylhydrazine as N Source, *Jpn. J. Appl. Phys.*, Vol. 46, pp. 2844-2847.

Ohshita Y, N. Kojima T. Tanaka T. HondaM. Inagaki and M. Yamaguchi Novel material for super high efficiency multi-junction solar cells *J. Cryst. Growth*, Vol 38, No 1, (March 2011) pp 328-331.

Pavelescu E.-M., Wagner J., Komsa H.-P., Rantala T. T., Dumitrescu M.& Pessa M., Nitrogen incorporation into GaInNAs lattice-matched to GaAs: The effects of growth temperature and thermal annealing, *J. Appl. Phys.*, 2005, Vol.78, No 8, pp. 083524-1-4.

People R.& Bean J. C. (1985) Calculation of critical layer thickness versus lattice mismatch for Ge_xSi_{1-x}/Si strained-layer heterostructures, *Appl. Phys. Lett.*, Vol. 47, No 3, pp. 322-324.

Ptak A. J., Friedman D. J., Kurtz S., & Keihl J.(2005). Enhanced depletion width GaInNs solar cells grown by molecular beam epitaxy, *Proceedings of 31IEEE PVSC*, pp. 603–606, Orlando, Florida, USA, January 3-7, 2005.

Reynolds C.L. and Tamargo M.C. (1984) LPE appartus with improved thermal geometry, US Patent 4 470 368.

Scheel H.J.(2003) Control of Epitaxial Growth Modes for High-performance Devices, In: *Crystal Growth Technology*, Scheel H.J.& Fukuda T., pp. 623-642, John Wiley & Sons, Ltd.

Shan W, Walukiewicz W, Ager JW, Haller EE, Geisz JF, Friedman DJ, Olson JM, Kurtz SR. (1999) Band Anticrossing in GaInNAs Alloys. *Phys. Rev. Lett.*, Vol. 82, No 6, pp 1221–1224.

Sheldon P., Jones K.M., Al-Jassim M.M., & Yacobi B.G.(1988) Dislocation density reduction through annihilation in lattice-mismatched semiconductors grown by molecular beam epitaxy, *J. Appl. Phys.*, Vol. 63, No 11, pp. 5609-5611.

Skierbiszewski C., Perlin P., Wisniewski P., Knap W., Suski T., Walukiewicz W., Shan W., Yu K. M., Ager J. W., Haller E. E., Geisz J. F., and Olson J. M. (2000) Large, nitrogen-induced increase of the electron effective mass in $In_yGa_{1-y}N_xAs_{1-x}$, *Appl. Phys. Lett.* Vol 76, No 17, pp 2409-2411.

Spruytte S. G., Coldren C. W., Harris J. S., Wampler W., Krispin P., Ploog K., & Larson M. C. (2001) Incorporation of nitrogen in nitride-arsenides: Origin of improved luminescence efficiency after anneal, *J. Appl. Phys.*, Vol. 89, No 8 , pp. 4401-4406.

Stranski I. N. and Krastanov L. V. (1939). *Abhandlungen der Mathematisch-Naturwissenschaft lichen Klasse.*, Akademie der Wissenschaften und der Literatur in Mainz, Vol. 146, p. 797.

Talwar D. N. (2007) Chemical bonding of nitrogen in dilute InAsN and high In-content GaInAsN, *Phys. Stat. Sol. (c)* Vol. 4, No. 2, pp. 674– 677.

Van der Merwe J.H (1979), Critical Reviews in Solid State and Materials Science, in *Chemistry and Physics of Solid Surfaces*, editor R. Vanselow, CRC Press, Boca Raton, 1979, 209 p.

Vitanov P., Milanova M., Arnaudov B., Evtimova S., Goranova E., Koleva G., Bakardjieva Vl. & Popov G. (2010) Study of melt-grown GaAsN and InGaAsN epitaxial layers (2010). *Journal ofPhysics:ConferenceSeries*, Vol. 253, pp. 012045-1-6.

Volmer M. & Weber A.,(1926). Keimbildung in übersättigten Gebilden, *Z. Physik Chem.*, Vol.119, pp 277-301.

Walukiewicz W., (1996) Amphoteric native defects in semiconductors, *Appl. Phys. Lett.*, Vol. 54, No 21, pp. 2094-2096.

Weyeres M., Sato M., & Ando H.(1992) Red shift of photoluminescence and absorption in dilute gaasn alloy layers, *Japanese Journal of Applied Physics*, Vol. 31, pp. L853.

Yamaguchi M., Nishimura K., Sasaki T., Suzuki H., Arafune K., Kojima N., Ohsita Y., Okada Y., Yamamoto A., Takamoto T. & Araki K. (2008). Novel materials for high-efficiency III-V multi-junction solar cells, *Solar Energy*, Vol. 82, No 2, pp. 173-180.

Yu K. M., Walukiewicz W., Shan W., J. Wu, Ager J. W., HallerE. E., Geisz J. F., & Ridgway M. C. (2000a). Nitrogen-induced enhancement of the free electron concentration in sulfur implanted GaN_xAs_{1-x}, *Appl. Phys. Lett.* Vol. 77, No 18, pp 2858-2860.

Yu K. M., Walukiewicz, W., Shan W., Ager J. W., Wu J., Haller E. E., Geisz J. F., Friedman D. J. and Olson J. M. (2000b) Nitrogen-induced increase of the maximum electron concentration in group III-N-V alloys. *Phys. Rev. B*, Vol. 61, No 20, pp.R13337-R13340.

Zhang S. B.& Wei S. H. (2001). Nitrogen Solubility and Induced Defect Complexes in Epitaxial GaAs:N, *Phys. Rev. Lett.*, Vol. 86, No 9, pp. 1789–1792.

6

Relation Between Nanomorphology and Performance of Polymer-Based Solar Cells

Almantas Pivrikas[1,2]
[1]Physical Chemistry, Linz Institute for Organic Solar Cells,
Johannes Kepler University Linz
[2]School of Chemistry and Molecular Biosciences, Centre for Organic Photonics and
Electronics, The University of Queensland, Brisbane
Austria

1. Introduction

Global warming and climate change has sparked the interest in alternative energy sources.(Cox et al., 2000) Although solar power reaching the surface of the Earth is able to meet the demands of humanity at the present,(Turner, 1999) an important question remains: how to convert the solar power into electrical power efficiently and at low costs.(Glaser, 1968) Polymer-based organic solar cells offer a possible solution for low cost photovoltaic energy conversion.(Wohrle & Meissner, 1991) In general, organic electronics has created an immense academic interest due to unlimited and flexible molecular engineering possibilities, allowing new organic materials with tailored physical properties to be synthesized.(Forrest, 2005b)

The most promising aspect of organic solar cells is their potential economic advantage due to large-scale production posibilities using continuous and large scale roll-to-rool printing and coating techniques allowing to deposit the active film, electrodes, sealing layers, antireflecting coatings and other components on flexible substrates all-at-once.(Krebs, 2009) Various aesthetic form factors, usually considered to be important for the solar panel integration into buildings can be achieved with this type of solar cell. Desired device form, color and a wide range of applications including solar power stations, roof-tops, portable devices, textile integrated power supplies and other consumer products can be envisioned.

The relation between fabrication costs of photovoltaic modules and power conversion efficiency defines the market success, therefore both factors have to be considered from academic and industrial perspective.(Brabec, 2004) Due to low dielectric constants and weak van der Waals forces binding the organic molecules into a solid, excitons (electron and hole pair strongly bounded by Coulomb attraction) are the primary photoexcitations in organic solids.(Schwoerer & Wolf, 2007) In order to achieve high power conversion efficiency of organic solar cells excitons have to be separated into mobile charge carriers for photocurrent generation.(Forrest, 2005a) The bulk-heterojunction concept is employed to overcome the short exciton diffusion distance. The photoactive film of heterojunction is formed from the donor and acceptor materials which are phase-separated on the nanometer length scale, to facilitate the photo-induced charge transfer as well as create a percolating pathways for charge transport to the electrodes.(Brabec et al., 2001; Halls et al., 1995) Therefore, the nanomorphology of polymeric solar cells plays a crucial role for the performance of the devices.

Historically, thermal annealing of the film has been used to induce the phase separation between donor and acceptor in bulk-heterojunction blends.(Padinger et al., 2003) However, thermal treatment creates an additional fabrication step in the whole device fabrication process.   Later, various methods have been tested and employed to control the nanomorphology of the blends, namely use of solvents with different boiling points (choice of solvent), reduction of drying speed (rate of drying and vapor annealing), changing the solubility of materials, melting of bilayers and the use of processing additives.(Pivrikas et al., 2010b) The later method has received great academic interest as it removes the need for post-production treatment while at the same time allowing fine control of the nanomorphology in various donor-acceptor blends.(Lee et al., 2008)

In this work the factors limiting the power conversion efficiency of excitonic polymer-based bulk-heterojunction solar cells are discussed.   Various methods allowing the film nanomorphology to be controlled are reviewed. The use of processing additives to control the phase separation for the formation of an interpenetrating network and how this impacts the power conversion efficiency is described.

2. Excitonic polymer-based solar cells

Polymer-based bulk-heterojunction solar cells (BHSC) have already shown certified efficiencies above 8 % demonstrating ability to compete with inorganic solar cell systems (eg. amorphous silicon cells fabricated on flexible substrates). Efficiencies exceeding 10% for solution processed solar cells are expected to be achieved soon.(Nayak et al., 2011) The power conversion efficiency of BHSCs is determined by the photophysical processes under operational conditions. A fundamental understanding of the relation between light absorption charge separation, charge transport, recombination, and film nanostructure as well as between the various thin film fabrication and processing parameters (such as solvent composition, solution concentration, deposition atmosphere and process temperature) is needed for further improvements. These important parameters can be controlled to some extent by adjusting the required film composition or device structure.(Gunes et al., 2007)

2.1 Solar cell device structure

The typical device structure of most organic optoelectronic devices, including organic light-emitting diodes and solar cells, is shown in Fig. 1. The front electrode is based on a transparent conducting oxide, such as indium tin oxide (ITO), that serves as the high-work-function, positive electrode.(Brabec et al., 2001) To further improve the quality of the ITO electrode and aid hole (positive charge carrier) extraction from the active film, poly(3,4-ethylenedioxythiophene)-poly(styrene sulfonate) (PEDOT-PSS) layer (tens of nanometers thick) is coated on top, forming a smooth surface which is essential in thin film devices. The photoactive film, the donor and acceptor bulk-heterojunction blend is deposited on top of the PEDOT-PSS layer. The whole device is finished by thermally evaporating the back contact (negative electrode) under high vacuum. To achieve the built-in electric field needed for most devices to operate, the back electrode must be made from a low-work-function metal that serves as the negative electrode. In the operation of a typical polymer-fullerene bulk-heterojunction solar cell, electrons generated in the active layer are collected by the back electrode (anode), and holes are collected at the opposite electrode (cathode).

2.2 Solar cell fabrication techniques

While academic research is highly concentrated on improving the power conversion efficiency, there are other important aspects needed for commercial success, such as cell stability, degradation, low manufacturing costs with rapid large scale production. This has been summarized by the Venn diagram as the unification of challenges when trying to combine power conversion efficiency, processability and stability into final devices.(Jorgensen et al., 2008)

Solution processing is attractive for fabricating organic optoelectronic devices mainly due to its simplicity and applicability for large scale and low-cost production. Thin films can be formed in various ways: a) printing techniques including screen printing, pad printing, gravure printing, flexographic printing and offset printing; b) coating techniques including pin coating, doctor blading, casting, painting, spray coating, slot-die coating, curtain coating, slide coating and knife-over-edge coating. The only technique that in both categories is inkjet printing.

Spin coating has been the most common technique for polymeric solar cell fabrication with numerous reviews and fundamental studies available.(Norrman et al., 2005) This technique, widely used in the microelectronics industry to deposit photoresist on silicon wafers, allows for the reproducible formation of highly homogeneous films over large areas. A typical spin coating process involves application of a solution (with the organic semiconductors dissolved in a solvent) to a substrate which is then either accelerated to the required angular velocity or is already spinning at it, Fig. 2.(Krebs, 2009) A large portion of the solution is wasted leaving a thin film on the substrate. Film thickness, morphology and surface topography strongly depend on the rotational speed, viscosity, volatility, diffusivity, molecular weight and concentration of the solutes and solvents used.(Cohen & Gutoff, 1992)

2.3 Current-Voltage dependence of solar cells

The most important figure of merit describing the performance of a solar cell is the power conversion efficiency, which is determined from the current voltage characteristics of the solar cells under operational conditions. Typical current-voltage characteristics of solar cells under illumination is shown in Fig. 3.(Deibel & Dyakonov, 2010)

The accurate measurement of the PCE according to international standards has been described in the literature,(Shrotriya et al., 2006) and is eesential for reproducibility and comparison of results between different laboratories. The Shockley diode equation describes the

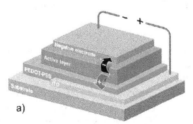

a)

Fig. 1. Schematic sandwich-type structure of organic solar cells showing an organic semiconductor active film between two metal electrodes with different work functions (typically ITO/PEDOT-PSS as positive and Ca/Al as negative contacts). Reprinted with permission from (Shaheen, 2007). Copyright 2007 Society of Photo-Optical Instrumentation Engineers.

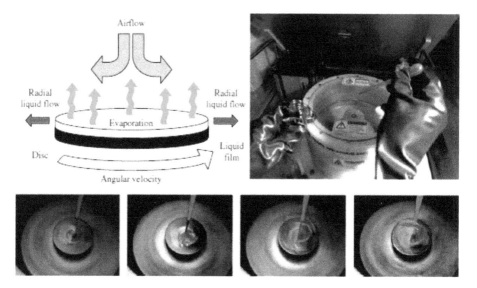

Fig. 2. Spin coating of organic solar cells from solution. Reprinted with permission from (Krebs, 2009). Copyright 2009, with permission from Elsevier.

current-voltage dependence of an ideal diode.(Shockley & Queisser, 1961) In the dark under forward bias the injection current increases exponentially with applied bias, whereas under reverse bias, current saturates at low applied voltages due to blocking contacts. This leads to a rectifying behaviour as can be clearly seen in the log-lin plot in Fig. 3. A description of the non-ideal device (typical organic solar cells) requires addition of series (R_s) and paralell (R_p) resistances. R_s is connected in series with the ideal diode and it describes the contact resistances such as injection barriers and sheet resistances. R_p arises due to the influence of local shunts between the two electrodes, i.e. additional current paths circumventing the diode. Typically in organic solar cells, a strong photocurrent dependence of applied electric field is observed manifesting as a non-saturated current at -1V reverse bias (Fig. 3). The field-dependent photocurrent arises due to:

a) field dependent mobile charge carrier generation, since an exciton has to dissociate into mobile carriers;(Oesterbacka et al., 2010)

b) charge carrier collection due to Hecht's law, if the extracted charge saturates at electric fields wheere the film thickness is larger than the carrier drift distance.(Hecht, 1932)

c) electric field dependent carrier mobility.(Pivrikas, Ullah, Sitter & Sariciftci, 2011)

In addition to electric field dependent mobility, the charge transport in disordered organic solar cells is also carrier concentration dependent. This effect arises due to the hopping nature of charge transport, where at higher carrier concentrations the carrier hopping probability between localized states increases (loosely speaking due to a higher density of localized states resulting in better electron wavefunction overlap) and therefore the carrier mobility increases.(PIVRIKAS, ULLAH, SINGH, SIMBRUNNER, MATT, SITTER & SARICIFTCI, 2011)

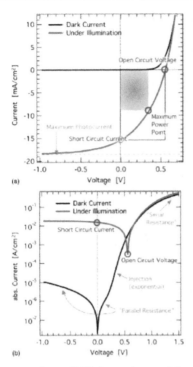

Fig. 3. Typical schematic current-voltage (I-V) dependence of the organic solar cell in the dark and under illumination. Reprinted with permission from (Deibel & Dyakonov, 2010). Copyright 2010, with permission from Institute of Physics.

2.4 Power conversion efficiency

PCE is defined as the ratio of the electrical power produced by a solar cell to the optical power of incident light (P_{light}):(Luque & Hegedus, 2003)

$$PCE = \frac{j_{SC}V_{OC}FF}{P_{light}} \tag{1}$$

where j_{SC} is the short circuit current density, V_{OC} is the open circuit voltage, $FF = V_{MP}j_{mp}/V_{OC}j_{SC}$ is the fill factor representing the maximum area in the fourth quadrant of the I-V characteristics of solar cells, V_{MP} and jmp are the voltage and current, respectively, at the point of maximum power, and P_{light} is the power of incoming light under Standard Test Conditions (1000 W/m2, AM 1.5 (Air Mass) solar reference spectrum, temperature during measurements 25 C).[15] To maximize the PCE values, all these parameters have to be maximized. The open circuit voltage of BHSC is determined by the energy of the quasi-Fermi levels of both semiconductors (donor and acceptor) as well as the Fermi levels of the electrodes.(Scharber et al., 2006) The short circuit current j_{SC} depends on electrical carrier drift (induced by electric field) and diffusion (induced by concentration gradient). Apart from the absorption on the film, charge carrier concentration and mobility are the main factors influencing the photocurrent regardless of the transport mechanism (drift or diffusion).(Nelson, 2003)

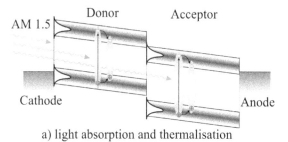

a) light absorption and thermalisation

b) exciton diffusion, dissociation and recombination

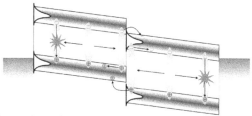

c) charge transport and recombination

Fig. 4. Factors limiting power conversion efficiency of excitonic solar cells. Reprinted with permission from (Pivrikas, 2010). Copyright 2010 IEEE.

The PCE of organic solar cells is influenced by many different photophysical processes and parameters. Simplified energy-level diagrams of organic solar cells utilizing donor and acceptor materials is shown in Fig. 2.(Pivrikas et al., 2010a)

2.5 Power conversion efficiency limiting mechanisms in excitonic solar cells

The first factor limiting the PCE is the absorption of light in the film. Ideally, as much as possible of the incident solar irradiance should be absorbed.(Pivrikas et al., 2010a) The Beer-Lambert law determines the light absorption profile in homogeneously distributed and scatter-free medium. Optical interference effects can also influence the light absorption profile in thin multilayer films.(Dennler et al., 2009) As can be seen in Fig. 4 some part of absorbed light energy is lost due to a thermalization process - charge relaxtion within the Density of States (DOS) to the lower energy levels to form an occupational-DOS within localized DOS.(Bassler, 1993; Juska et al., 2003; Osterbacka et al., 2003)

The second efficiency limiting process is exciton dissociation into mobile charge carriers. Due to low dielectric constants and consequently weak Coulombic field screening in the organic materials (relative static permittivity is around 3) the primary photoexcitation is an exciton, which does not create the photocurrent.(Luque & Hegedus, 2003) The exciton diffusion length describes how far an exciton can diffuse within its lifetime. The concept of heterojunction between two organic semiconductors (donor and acceptor) is used to split an exciton into mobile charge carriers. Efficient exciton dissociation (charge transfer) takes place at the interface between donor and acceptor if a suitable offset in energy level exists, as shown in Fig. 4 b).(Pivrikas et al., 2010a) The excited state Charge Transfer (CT) complex (sometimes called exciplex) might be formed after the dissociation of an exciton meaning that positive and negative charge carriers might remain bounded by the Coulomb attraction at the donor acceptor interface, which would not contribute to photocurrent.

The charge carrier transport (collection) to the electrodes is the third important limiting processes. As shown in Fig. 4 c), mobile electrons and holes must be transported to the opposite electrodes. The driving force can be either diffusion, related to the carrier concentration gradients, and/or drift due to a built-in electric field. An important aspects of charge transport are the charge extraction at the semiconductor-metal interface. The energy level alignment between the metal and semiconductor, free charge carrier concentration in the film (doping level) as well as the trapping level concetration, carrier capture and release times from capture centers determine the interfacial properties of the device. Non-blocking contact without energetical barrier for charge carrier extraction is required to be present at the interface.(Baranovski, 2006) The disordered nature of solution processed films of organic semiconductors results in low charge carrier mobilities (tzpically 10^{-3} - 10^{-7} $cm^2V^{-1}s^{-1}$ in π-conjugated polymers). The mobility of the slower charge carrier limits the photocurrent, and therefore the efficiency of the solar cell due to accumulation of charge carriers. Since the photocurrent under operation conditions typically approach space charge limited current, second order recombination processes become dominant due to high charge carrier concentration, and the carrier lifetime becomes shorter than the transit time.(Pivrikas et al., 2010a)

2.6 Bulk-heterojunction solar cells

The light absorption coefficient, α, in a disordered organic film is usually high, on the order of 10^5 cm^{-1}. This allows thin films, on the order of hundreds of nanometers, to be used in solar cells. However, the exciton diffusion length in most organic materials is of the order of 10 nm. If the exciton is to diffuse to the interface between the two materials (donor-acceptor) in order to separate into mobile charge carriers, these two materials must be blended on this length scale. Furthermore, the donor-acceptor phases must for bi-continuous network with percolating pathways for electron and hole transport to the elctrodes. This is the operating principle of the BHSC shown in Fig. 5.(Sariciftci, 2006) The film nano-morphology is crucially important for the efficiency of solar cells.(Ma et al., 2005) The nanoscale phase-separation phase separation between donor and acceptor in BHSC plays an important role relating the device properties and performance to the solar cell fabrication methods. Typical donor is poly(3-hexylthiophene-2,5-diyl) (P3HT) and acceptor is [6,6]-phenyl-C61-butyric acid methyl ester (PCBM)

3. Methods to control the morphology of BHSC

The formation and the size of nanoscale domains of donor and acceptor phases are strongly dependent on the film fabrication techniques and conditions. Beyond the selection of suitable materials there are several parameters that must be carefully controlled when fabricating

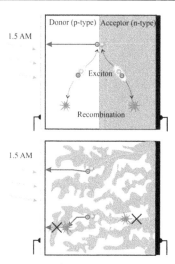

Fig. 5. Bilayer and bulk-heterojunction solar cells. Reprinted with permission from (Pivrikas et al., 2010a). Copyright 2009, with permission from Elsevier.

BHSC, such as the solution concentration, deposition temperature, donor-acceptor blend ratio, spin speeds using solvents with different boiling points, solvent evaporation kinetics, vapour pressure, solubility, and polarity.(Pivrikas et al., 2010b)

Various methods allowing the control of the film nano-morphology have been introduced in the past.(Chen, Hong, Li & Yang, 2009; Peet et al., 2009) Initially it was observed that the PCE of BHSC significantly increases upon a postproduction treatment, e.g. thermal annealing of solar cells with applied external voltage.(Hoppea & Sariciftci, 2004; Padinger et al., 2003) Other methods used in the past to control the morphology of BHSC involve the use of appropriate solvents with specific boiling point that allow either the increase or decrease of the solvent evaporation rate.(Kim, Choulis, Nelson, Bradley, Cook & Durrant, 2005; Shaheen et al., 2001; Yu et al., 1995) Other methods, such as reducing the drying speed of spin-coated films,(Li et al., 2005; Mihailetchi et al., 2006; Vanlaeke et al., 2006) solubility matching(Troshin et al., 2009) and the melting of bilayers have also been used.(Kim, Liu & Carroll, 2006) It was observed that chemical additives can substitute the post production treatment of BHSC.(Lee et al., 2008) Processing additives are an attractive concept due to the simplicity and suitability for large scale production.

3.1 Thermal annealing of devices

Thermal annealing, by controlling the temperature and annealing time, is typically applied to either the final device or BHJ films in order to improve the nanoscale phase separation between donor and acceptor.(Ma et al., 2005; Sun et al., 2007; Xin et al., 2008; Zhang, Choi, Haliburton, Cleveland, Li, Sun, Ledbetter & Bonner, 2006) Significant improvement in photovoltaic performance after annealing is typically observed in P3HT/PCBM blends.(Padinger et al., 2003) Thermal annealing has the advantage in that it can be applied independently of the film deposition technique. Thermal annealing has also been shown to enhance the crystallinity of the polymer, such as for P3HT, increasing the PCE and the photocurrent due to increased carrier mobility.(Erb et al., 2005) Furthermore, the interconnections between the polymer/fullerene phazes in the interpenetrating network are enhanced as a result of phase separation between the donor and acceptor on meso (> 100 nm) and nanoscales (< 20 nm).(Kim, Cook, Tuladhar, Choulis, Nelson, Durrant, Bradley, Giles, McCulloch, Ha et al., 2006; Ma et al., 2005)

The effect of thermal annealing on film morhology was clearly demonstrated by bright-field (BF) transmission electron microscopy (TEM) images, recorded in slight defocussing conditions in a P3HT/PCBM blends, as shown in Fig. 6.(Yang et al., 2005)

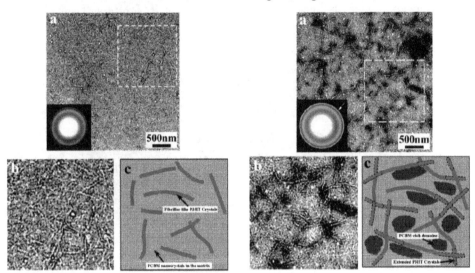

Fig. 6. Transmission Electron Microscopy (TEM) images show the overview (a) and zoomed in (b), and the corresponding schematic representation (c) of the photoactive layer solar cells. Left (a)-(c) images: pristine unannealed P3HT and PCBM blend. Right (a)-(c) images: thermal annealed P3HT and PCBM blend. Insets in (a) figures is the corresponding SAED pattern. The dash line bordered regions represent the extension of existing P3HT crystals in the pristine film or newly developed PCBM-rich domains during the annealing step. The arrow is to indicate the increased intensity of (020) Debye-Scherrer ring from P3HT crystals compared to the SAED pattern shown in the inset of Figure 2a. Reprinted with permission from (Yang et al., 2005). Copyright 2005 American Chemical Society.

As can be seen, fibrillar-like P3HT domains (brighter in contrast compared to background) overlap with each other over the whole film. The inset of Fig. 6 shows the selected area electron diffraction (SAED) pattern of the film. The outer ring corresponds to a distance of 0.39 nm, which is typical pi-pi stacking distance of P3HT chains.(Ihn et al., 1993) The crystallinity of P3HT crystals is not very pronounced as seen from the low intensity of the reflection ring. The inner ring in the SAED pattern, corresponding to a d-spacing of 0.46 nm, is seen to be even more diffuse and has been attributed by the nanometer sized PCBM nanocrystals that are homogeneously dispersed throughout the film. Fig. 6 a) shows the BF TEM images of the composite film after annealing (120 C for 60 min). The most pronounced feature in the BF TEM image of the annealed sample is the increased contrast and the appearance of bright fibrillar P3HT crystals throughout the entire film. The width of these crystals remains almost constant compared to the pristine composite film, but on average their length was found to increase over 50 %. The increased crystallinity of P3HT after thermal treatment is visible from the increased intensity of the reflection ring in the SAED pattern (inset of Fig. 6 a)). Larger and darker PCBM rich areas can be observed suggesting an increased phase demixing between P3HT and PCBM. It was concluded that the crystallinity of P3HT is improved upon annealing and the demixing between the two components is increased, but large-scale phase separation does not occur. The resulting interpenetrating networks composed of P3HT crystals with a high aspect ratio and aggregated nanocrystalline PCBM domains provide continuous pathways in the entire photoactive layer for efficient hole and electron transport.

In order to further understand the extent of thermal annealing, 2-D X-ray scattering in a grazing incidence geometry (GIWAXS) was used to study the development of the crystalline structure of P3HT and PCBM during the interdiffusion process at various temperatures. 2D GIWAXS patterns of as-prepared P3HT/PCBM bilayers and annealed samples are shown in Fig. 7.(Treat et al., 2011) The diffraction patterns for P3HT shows that the *a*-axis of the P3HT crystals is predominantly oriented perpendicular to the substrate and the *b*-axis (pi-stacking)

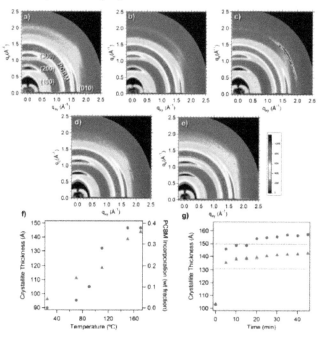

Fig. 7. Two-dimensional GIWAXS of a P3HT/PCBM bilayer on Si a) as-cast and annealed at b) 70 C, c) 110 C, d) 150 C, and e) 170 C for 5 min. f) The Scherrer equation was used to extract the P3HT crystallite thickness along the a- axis from the full-width-at-half-maximum of the (100) reflection. PCBM incorporation from the DSIMS measurements was plotted for comparison at various annealing temperatures. g) Growth in the crystal thickness with time using in-situ heating 2D GIWAXS of a P3HT/PCBM bilayer on Si at 110 C (orange) and 170 C (blue). The Scherrer equation was used to determine crystal thickness from the (100) reflection corresponding to P3HT. The dotted line corresponds to a neat P3HT/Si sample heated for 5 min at 110 C (orange) and 170 C (blue). Copyright 2011 Wiley. Used with permission from (Treat et al., 2011).

is oriented parallel - which is the typically observed P3HT orientation. Upon annealing the as-prepared films at various temperatures, the d-spacing along the a-axis of the P3HT crystal was found to remain constant, indicating that during the interdiffusion process, the PCBM does not interpenetrate between the side chains of the P3HT crystal structure.(Mayer et al., 2009) The peak width of the diffraction ring, corresponding to the aggregates of PCBM does not change during the interdiffusion process, showing that PCBM remains in an amorphous state with aggregates large enough to scatter incident X-rays. Only a small change in the distribution of P3HT crystal orientations was found to be present at various levels of interdiffusion, while the intensity of the (200) peak of P3HT increased by nearly a factor of two on annealing at 170 C. It was shown that the interdiffusion process has little effect on the crystalline regions of the P3HT film, where the diffusion of PCBM into P3HT occurs within the disordered regions of P3HT.

To determine how interdiffusion within this system affects the growth of the P3HT crystallites, the P3HT crystallite size along the a-axis for the bilayer films was compared to pure P3HT films heated under similar conditions (Fig. 7 (f)-(g)). The P3HT crystallite size was estimated using the Scherrer equation and plotted against the fraction of PCBM within the P3HT layer (Fig. 7 (f)). The crystallite size was found to increase with increasing annealing temperature regardless of the level of interdiffusion. The P3HT crystallite size in the bilayer system was found to increase most rapidly during the first 5 min of annealing, where the crystallite thickness was approching that for a neat P3HT film heated under similar conditions (Fig. 7 (g)).

3.2 Solvent effects

Postproduction treatment requires a rather well controlled environment, it adds an additional fabrication costs to the solar cell manufacturing process, which might not be attractive for large-scale industrial production. Furthermore, some material systems, like the low band gap organic semiconductor poly[2,6-(4,4-bis-(2-ethylhexyl)-4H-cyclopenta[2,1-b;3,4-b0]-dithiophene)-alt-4,7-(2,1,3-benzothiadiazole)] (PCPDTBT) blended with [6,6]-phenyl C71-butyric acid methyl ester (C71-PCBM), do not shown any improvement upon thermal annealing.

Phase separation and molecular self-organization can be influenced by solvent evaporation since the solvent establishes the film evolution environment. Slow drying or solvent annealing techniques have also been used to control the morphology of the blends by changing the rate of solvent removal.(Li et al., 2005; Li, Yao, Yang, Shrotriya, Yang & Yang, 2007; Sivula et al., 2006) The use of different solvents and their effect on the film nano-structure of BHSC has been studied in detail in the past.(Li, Shrotriya, Yao, Huang & Yang, 2007) High boiling point solvents were used with the device placed in an enclosed container, in which the atmosphere rapidly saturates with the solvent.

Grazing-incidence x-ray diffraction (GIXRD) studies provided evidence that the solvent evaporation rate directly influences the polymer chain arrangement in the film.(Chu et al., 2008) It was shown that the use of higher boiling point solvent strongly improves the PCE of MDMO-PPV and PCBM blends.(Shaheen et al., 2001) Higher PCE values due to improved film morphology and crystallinity have been reached by substituting chloroform with chlorobenzene for P3HT/PCBM BHSC.(Ma et al., 2005) The difference between chlorobenzene and 1,2-dichloro benzene for use as a solvent was shown in the novel low bandgap polymer PFco-DTB and C71-PCBM blend systems, where chlorobenzene resulted in films with higher

roughness.(Yao et al., 2006) Non-aromatic solvents have shown to be able to affect the photovoltaic performance of MEH-PPV and PCBM blends.(Yang et al., 2003)
An interesting method to study the morphology of BHSC optically by recording exciton lifetime images within the photoactive layer of P3HT and PCBM has been demonstrated by Huan et al.(Huang et al., 2010) Using a confocal optical microscopy combined with a fluorescence module they were able to image the spacial distrubution of exciton lifetime for both slow and fast dried films, as shown in Fig. 8.

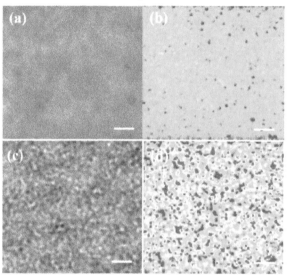

Fig. 8. (a, c) Transmitted images and (b, d) exciton lifetime images of the BHJ film prepared from rapidly and slowly grown methods, respectively, measured after excitation at 470 nm using a picosecond laser microscope (512 × 512 pixels). Scale bars: 2 μm. Reprinted with permission from (Huang et al., 2010). Copyright 2010 American Chemical Society.

The transmitted image of the rapidly grown film (Fig. 8 (a)) shows a uniform and featureless characteristics throughout the structure, indicating that P3HT and PCBM were mixed well within the films. This monotonous transmitted image corresponds to a uniform exciton lifetime distribution. Fig. 8 (c)-(d) shows transmitted and exciton lifetime images for the slowly dried films. The bright spots are emissions from many polymer chains that have stacked or aggregated into a bulk cluster leading to a reduced PL quenching. The red regions (P3HT-rich domains Fig. 8 (d)) correspond to the bright spot of the transmitted image (Fig. 8 (c)). In agreement with previous studies, the images showed that the active layers during slow solvent evaporation provide a 3D pathways for charge transport reflecting better cell performance.

3.3 Processing additives

This method is based on the usage of a third non-reacting chemical compound, a processing additive, to the donor and acceptor solution. Improvement of the performance of polymer/fullerene photovoltaic cells doped with triplephenylamine has been reported.(Peet et al., 2009) The ionic solid electrolyte (LiCF3SO3) used as a dopant also resulted in enhanced PCE of MEH-PPV/PCBM blends due to an optimized polymer morphology, improved electrical conductivity and in situ photodoping.(Chen et al., 2004) A copolymer including thieno-thiophene units (DHPT3) has been used as a nucleating agent for crystallization in the active layer of P3HT and PCBM BHSC.(Bechara et al., 2008) It was demonstrated that the addition of DHPT3 in P3HT/PCBM thin films induces a structural ordering of the polythiophene phase, leading to improved charge carrier transport properties and stronger active layer absorption. High-performance P3HT/PCBM blends were fabricated using quick drying process and 1-dodecanethiol as an additive.(Ouyang & Xia, 2009) Ternary blends of P3HT, PCBM and poly(9,9-dioctylfluorene-co-benzothiadiazode) (F8BT) showed enhanced optical absorption and partly improved charge collection.(Kim, Cook, Choulis, Nelson, Durrant & Bradley, 2005) A few volume percent of 1,8-diiodooctane in o-xylene was used to dissolve poly(9,9-di-n-octylfluorene) PFO allowing the control of film morphology.(Peet et al., 2008) Block-copolymers and diblock copolymers with functionalized blocks have also shown to be able to influence the film morphology.(Sivula et al., 2006; Sun et al., 2007; Zhang, Choi, Haliburton, Cleveland, Li, Sun, Ledbetter & Bonner, 2006)

3.3.0.1 "Bad" solvent effect

The incorporation of other solvents into the host solvent is capable of controlling the film morphology of BHSC.(Chen et al., 2008; Wienk et al., 2008; Xin et al., 2008; Zhang, Jespersen, Björström, Svensson, Andersson, Sundstr"om, Magnusson, Moons, Yartsev & Ingan"as, 2006) In some cases, changes in the solvent composition lead to interchain order that cannot be obtained by any other method.(Campbell et al., 2008; Moulee et al., 2008; Peet et al., 2007) The use of nitrobenzene as an additive has been shown to improve the phase-separation between the donor and acceptor (P3HT/PCBM blend), where P3HT was shown to be present in both amorphous and crystalline phase.(Moule & Meerholz, 2008; van Duren et al., 2004)

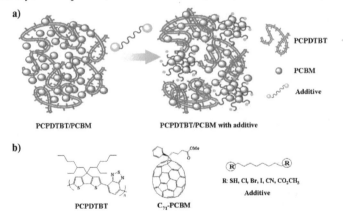

Fig. 9. Schematic depiction of the role of the processing additive in the self-assembly of bulk heterojunction blend materials (a) and structures of PCPDTBT, C71-PCBM, and additives (b). Reprinted with permission from (Lee et al., 2008). Copyright 2008 American Chemical Society.

The concept of mixing a host solvent with a "bad" solvent has been explored resulting in solvent-selection rules for desired film morphology.(Alargova et al., 2001) Solvents, distinctly dissolving one component of the blend, induce the aggregation of nanofibers and nanoparticles in the solvent prior to film deposition.(Yao et al., 2008) It was shown that (independent of the concentration of the additive) fullerene molecules crystallized into distributed aggregates in the presence of a "bad" solvent in the host solvent. Well aligned P3HT aggregates resulting in high degree of crystallinity due to the interchain $\pi - \pi$ stacking were observed upon addition of hexane.(Li et al., 2008; Rughooputh et al., 1987) The addition of 1-chloronaphthalene (a high boiling point solvent) into dichlorobenzene has also resulted in similar self-organization of polymer chains.(Chen et al., 2008) It was shown that in the blends of poly(2,7-(9,9-dioctyl-fluorene)-alt-5,5-(40,70-di-2-thienyl-20,10,3-benzothiadiazole)) and PCBM dissolved in chloroform with a small addition of chlorobenzene, a uniform domain distribution was attained, whereas the addition of xylene or toluene into the chloroform host solvent resulted in larger domains, stronger carrier recombination and a smaller photocurrent. Alkane-thiol based compounds were extensively used as processing additives in the past.(Lee et al., 2008) The photoconductivity response was shown to increase strongly in polymer/fullerene composites by adding a small amount of alkane-thiol based compound to the solution prior to the film deposition.(Coates et al., 2008; Peet et al., 2006) By incorporating a few volume percent of alkanethiols into the PCPDTBT/C71-PCBM BHSC (Fig. 9) it was shown that the PCE improves almost by a factor of two.(Alargova et al., 2001; Peet et al., 2007)

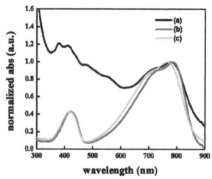

Fig. 10. UV-visible absorption spectra of PCPDTBT/C71-PCBM films processed with 1,8-octanedithiol: before removal of C71-PCBM with alkanedithiol (black); after removal of C71-PCBM with alkanedithiol (red) compared to the absorption spectrum of pristine PCPDTBT film (green). Reprinted with permission from (Lee et al., 2008). Copyright 2008 American Chemical Society.

The alkanedithiol effect was explained by the ability of alkanedithiols to selectively dissolve the fullerene component, where the polymer is less soluble, Fig. 9 The effect has been proven by removing the fullerene domains by dipping the BHJ film into an alkanedithiol solution and measuring light absorption before and after dipping.(Lee et al., 2008) The normalized absorption spectra (shown in Fig. 10) demonstrate that after dipping the film the absorption matches that of the pristine polymer.

As a consequence, "bad" solvent addition provides a means to select solvent-additives in order to control the phase-separation in BHSC. It was shown that during film processing the fullerene stays longer in its dissolved form, due to the rather high boiling point of alkanedithiol (> 160 C), allowing for self-aligning and phase-separation between the polymer and fullerene as suggested in Fig. 7 b). Two effects control the morphology of the blends:
a) selective solubility of one of the components;
b) a high boiling of the additive compared to the host solvent.

The concentration of the processing additive allows the amount of phase-separation between the donor and the acceptor to be controlled.

3.3.0.2 Different processing additives

1,8-di(R)octanes with various functional groups (R) allow control of the film morphology.(Peet et al., 2007) The best results were obtained with 1,8-diiodooctane. Progressively longer alkyl chains, namely 1,4-butanedithiol, 1,6-hexanedithiol, 1,8-octanedithiol or 1,9-nonanedithiol were used to manipulate the morphology of solution processed films. It was concluded that approximately six methylene units are required for the alkanedithiol to have an appreciable effect on the morphology.

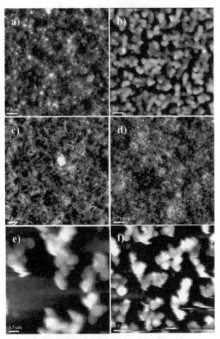

Fig. 11. AFM topography of films cast from PCPCTBT/C71-PCBM with additives: (a) 1,8-octanedithiol, (b) 1,8-cicholorooctane, (c) 1,8-dibromooctane, (d) 1,8-diiodooctane, (e) 1,8-dicyanooctane, and (f) 1,8-octanediacetate. Reprinted with permission from (Chen, Yang, Yang, Sista, Zadoyan, Li & Yang, 2009). Copyright 2009 American Chemical Society.

Fig. 11 shows a Atomic Force Microscopy (AFM) surface topography of films cast from PCPCTBT/C71-PCBM with the various processing additives.(Lee et al., 2008) The 1,8-octanedithiol (a), 1,8-dibromooctane (c), and 1,8-diiodooctane (d) resulted in phase-segregated morphologies with finer domain sizes than those obtained with 1,8-dichlorooctane (b), 1,8-dicyanooctane (e), and 1,8-octanediacetate (f). The morphology of films processed with 1,8-diiodooctane showed more elongated domains than those processed with 1,8-octanedithiol and 1,8-dibromooctane. The 1,8-di(R)octanes with SH, Br, and I, gave finer domain sizes and exhibited more efficient device performances than those with $R = Cl$, CN, and CO_2CH_3. The AFM images of the BHJ films processed using 1,8-di(R)octanes with

$R = Cl$, CN, and CO_2CH_3 showed large scale phase separation with round-shape domains and no indication of a bicontinuous network.

3.3.0.3 Concentration of processing additives

Once the most effective thiol functional group has been indentified, it is interesting to find how the concentration of the processing additive in solution affects the film morphology. The effect of additive concentration in the solution was clearly observed in surface topography images in AFM.(Chen, Yang, Yang, Sista, Zadoyan, Li & Yang, 2009)

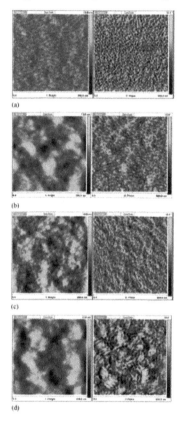

Fig. 12. Tapping mode AFM images of films with different amounts of 1,8-octanedithiol in 500 nm × 500 nm. Left: topography. Right: phase images. (a) 0 μL, (b) 7.5 μL, (c) 20 μL, and (d) 40 μL of 1,8-octanedithiol. The scale bars are 10.0 nm in the height images and 10.0 ° in the phase images. Reprinted with permission from from (Chen, Yang, Yang, Sista, Zadoyan, Li & Yang, 2009). Copyright 2009 American Chemical Society.

AFM images (a), (b), (c), and (d) of Fig. 12 show the height (left) and phase (right) images of polymer films with 0, 7.5, 20, and 40 μL of 1,8-octanedithiol, respectively, showing an increasing trend in roughness with increasing amount of 1,8-octanedithiol. The domain sizes were found to be consistent with the higher crystallization observed with increasing amount of 1,8-octanedithiol. Finely dispersed structures were observed when there was no

1,8-octanedithiol added. The AFM results were consistent with PL spectra showing higher PL intensity with increased 1,8-octanedithiol concentration.

AFM provides information about the film surface only, the bulk of the film has been studied using synchrotron-based grazing incidence X-ray diffraction (GIXD) in P3HT:PCBM blends.(Chen, Yang, Yang, Sista, Zadoyan, Li & Yang, 2009) Fig. 13 (a) represents 2-D GIXD

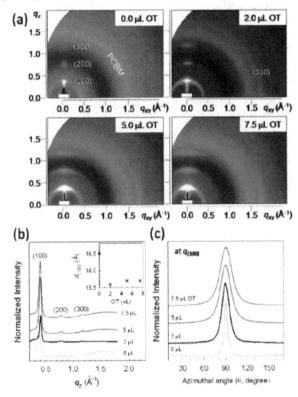

Fig. 13. (a) 2D GIXD patterns of films with different amounts of 1,8-octanedithiol. (b) 1D out-of-plane X-ray and (c) azimuthal scan (at $q^{(100)}$) profiles extracted from (a). Inset of b: calculated interlayer spacing in the (100) direction with various amounts of 1,8-octanedithiol. Reprinted with permission from (Chen, Yang, Yang, Sista, Zadoyan, Li & Yang, 2009). Copyright 2009 American Chemical Society.

patterns of the as-spun P3HT:PCBM films with different concentrations of 1,8-octanedithiol. It was found that the hexyl side chains and backbone of P3HT are oriented perpendicular and parallel to the surface, respectively regardless of 1,8-octanedithiol concentration. However, the crystallinity of P3HT in the films significantly increases in the presence of 1,8-octanedithiol and tends to keep steady above 5 μL 1,8-octanedithiol, as seen from in 1-D out-of-plane X-ray profiles normalized by film thicknesses (see Fig. 13 (b). The average interlayer spacing was observed to change significantly in the presence of 1,8-octanedithiol. It was concluded that the interaction between P3HT is stronger in the presence of 1,8-octanedithiol with the P3HT crystallinity improved due to stacking. The size distribution of P3HT crystals was found to be broader with increasing amount of 1,8-octanedithiol, as shown in Fig. 13 (c).

Improved crystallization of P3HT and broader crystal size distribution at higher 1,8-octanedithiol concentrations was explained by solvent volume ratios. During the film fabrication, the main solvent evaporates faster than the additive solvent resulting in a sudden increase of the volume ratio of the additive solvent to the main solvent. Polymer molecules lower their internal energy by aggregating when the additive solvent volume ratio reaches a critical point. At higher additive concentrations, the time required to reach this point is reduced and aggregation is stronger. As a result, polymer molecules aggregate with larger average domain sizes due to the stronger driving force and broader size distributions arises due to the shorter aggregation time.

4. Schematic structures of bulk-heterojunction film morphology

The morphological studies discussed above highlight the importance of phase separation between donor and acceptor, and reveal a schematic film structures for polymer-based bulk-heterojunction solar cells, as shown in Fig. 14..(Hoppe et al., 2006; Huang et al., 2010; Peumans et al., 2003)

In the top Fig. 14 (a), the percolated pathways for electrons and holes is created allowing them to reach the respective electrodes. In Fig. 14 b the situation for an enclosed PCBM cluster is shown: here electrons and holes will recombine, since percolation is insufficient.

The center Fig. 14 show that the lower surface energy of P3HT, relative to PCBM, provides the driving force for the interconcentration gradient observed in both the rapidly (a) and slowly (b) grown films. The film prepared through a rapidly grown process leads to an extremely homogeneous blends. A greater number of percolating pathways are formed in slow grown films.

Furthermore, the effect of annealing on the interface morphology of a mixed-layer device was modeled using a cellular model, as shown in Fig. 14 (bottom) for different temperatures. Annealing temperatures has been shown to crucially influence the morphology of the mixed-layer device, while the modeled morphology resemble experimentally measured devices.

5. Processing additive effect on solar cell performance

The photophysical effects of 1,8-octanedithiol (ODT) additives on PCPDTBT and C71-PCBM composites and device performance were studied using photo-induced absorption spectroscopy.(Hwang et al., 2008) Reduced carrier loss due to recombination was found in BHJ films processed using the additive. From photobleaching recovery measurements reduced carrier losses were demonstrated. However, it was concluded that the amount of the reduction is not sufficient to explain the observed increase in the power conversion efficiency (by a factor of 2). Carrier mobility measurements in Field Effect Transistor (FET) configuration demonstrated that the electron mobility increased in the PCPDTBT:C71-PCBM when ODT is used as an additive, resulting in enhanced connectivity of C71-PCBM networks.(Cho et al., 2008) This work also showed that if the ODT was not completely removed from the BHJ films by placing them in high vacuum ($> 10^{-6}$ torr) the hole mobility actually decreased, implying that residual ODT may act as a hole trap. It was concluded that the improved electron mobility was the primary cause of the improved power conversion efficiency, while the hole mobility was found to be relatively insensitive to the additive.

5.1 Power conversion efficiency and current-voltage dependence

In order to clarify the effect of chemical additives on the photophysical properties and photovoltaic performance, regioregular P3HT and PCBM bulk-heterojunction solar cells were fabricated in four different ways:
(1) as produced films (untreated, no alkyl thiol);
(2) thermally annealed films (refereed to as treated in text, no alkyl thiol);
(3) as produced films with alkyl thiol (refereed to as treated in text, with alkyl thiol);
(4) thermally annealed films with alkyl thiol (refereed to as treated in text, with alkyl thiol).
The fabrication procedures were kept the same for all four types of cells. The details on device preparation can be found elsewhere.(Pivrikas et al., 2008)
Current-voltage (I-V) characteristics under illumination of devices are shown in Fig. 15. Untreated solar cells gave the worst performance with the least short circuit current and low fill factor. However, these cells demonstrate a relatively higher open circuit voltage, but, due to a low short circuit current and a low fill factor, their power conversion efficiency was low, around 1 %. The difference in photocurrents between annealed cells and these with alkyl thiol

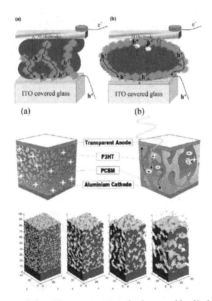

Fig. 14. Schematic structures of the film nanomorphology of bulk-heterojunction blends - all emphasizing the importance of the interpenetrating network in polymer-based solar cells. Top figures: (a) chlorobenzene and (b) toluene cast MDMO-PPV and PCBM blend layers. Center figures: vertical phase morphology of (a) rapidly and (b) slowly grown P3HT and PCBM blends. Bottom figures: the simulated effects of annealing on the interface morphology of a mixed-layer photovoltaic cell. The interface between donor and acceptor is shown as a green surface. Donor is shown in red and acceptor is transparent. Top figures reprinted with permission from (Hoppe et al., 2006), copyright 2006, with permission from Elsevier. Middle figures reprinted with permission from (Huang et al., 2010), copyright 2010 American Chemical Society. Bottom figures adapted by permission from Macmillan Publishers Ltd: (Peumans et al., 2003), copyright 2003.

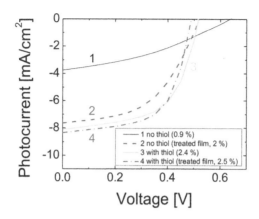

Fig. 15. Current-voltage characteristics demonstrating significant performance improvement under illumination (1000 W/m², 1.5 AM) for P3HT/PCBM bulk-heterojunction solar cells prepared in different ways: as produced (thin line), annealed (thick dashed line), thiol added (thick line), thiol added and annealed (thick dash dot line). Reprinted with permission from (Pivrikas et al., 2008). Copyright 2008, with permission from Elsevier.

is small, except that treated cells have lower fill factors and therefore slightly lower efficiency as compared to those with alkyl thiol additive, Fig. 16.

5.2 Light absorption and external quantum efficiency
In order to clarify the factors determining OPV device efficiency, the incident photon to current efficiency (IPCE), alternatively called External Quantum Efficiency (EQE) is measured, since it provides information on light absorption spectra, charge transport and recombination losses. The effect of thermal treatment versus processing addictive, as well as the effect of additive concentration, was studied and shown in Fig. 16. In Fig. 16 (a) and (d) the light absorption and Beer-Lambert absorption coefficient are shown as a function of wavelength. In agreement with previous observations, an increase in optical absorption is seen for treated cells. The red-shift of the absorption and characteristic vibronic shoulders are clearly pronounced in treated cells (at around 517 nm, 556 nm and 603 nm) both arising from strong interchain interactions within high degree of crystallinity in P3HT. In solution, no peak shift was observed, suggesting that the influence of the additive on P3HT happens during the solvent drying (or spin coating) process and not in the solution state. The increase in optical absorption at higher additive concentrations demonstrates that more energy can be harvested in solar cells, therefore, these cells have better photovoltaic performance due to a larger amount of photons being absorbed in the film.

While PCBM is known to quench the PL of P3HT effectively in the well mixed blends.(Chen, Yang, Yang, Sista, Zadoyan, Li & Yang, 2009) The photoluminescence was shown to increase with increasing amount of 1,8-octanedithiol (Fig. 16 (b)), suggesting that the phase separation between the P3HT and PCBM is increasing since the exciton diffusion distance is on the same order of magnitude.(Xu & Holdcroft, 1993; Zhokhavets et al., 2006)

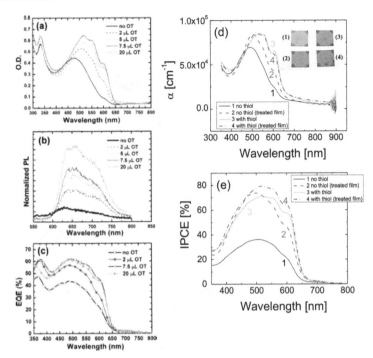

Fig. 16. Changes in light absorption (a) and photoluminescence (PL) (b) and External Quantum Efficiency (EQE) (c) shown at various amounts of processing additive (OT is 1,8-octanedithiol) used during film preparation. Changes in light absorption (d) and incident photon to current efficiency (IPCE) in (e) measured in pristine and treated (annealed films and films fabricated with processing additive) films. Strong red-shift in absorption, appearance of absorption peaks, higher IPCE values in treated films or films with processing additive well agrees with improved OPV performance. Thermal annealing of films fabricated with processing additive results in no change in OPV performance. Figures on the left reprinted with permission from (Chen, Yang, Yang, Sista, Zadoyan, Li & Yang, 2009). Copyright 2009 American Chemical Society. Figures on the right reprinted with permission from (Pivrikas et al., 2008). Copyright 2008, with permission from Elsevier.

A strong improvement in IPCE was observed in treated solar cells. The IPCE dependence approximately follows the light absorption curve, as the same characteristic absorption peaks are reproduced in the optical absorption spectra (Fig. 16). From the IPCE studies it was concluded that the improvement in the performance of solar cells is not only due to the increased optical absorption, but also due to improved transport (higher carrier mobility) and/or reduced recombination losses (eg. due to longer charge carrier lifetime), which again confirms the benefits of improved interpenetrating network between donor and acceptor.

5.3 Charge transport

Since it was found from ICPE studies that the film morphology not only improves the light absorption, but also results in better charge transport, it is important to quantify this improvement. In order to understand the difference in charge transport properties in treated and untreated cells, dark IV curves were recorded for all 4 types of treated cells shown in Fig. 17.(Pivrikas et al., 2008)

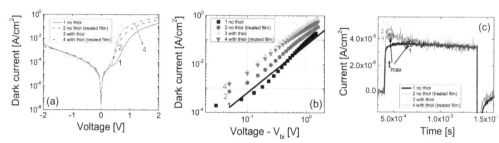

Fig. 17. The improvement in charge carrier mobility in treated (annealed films and films fabricated with processing additive) compared to pristine films demonstrated by two methods: dark current-voltage injection and CELIV. (a) log-lin plot showing the rectification ratio in forward and reverse bias and insignificant differences in leakage current in reverse bias. (b) log-log plot in forward bias showing much higher injection current levels in treated blends. (c) faster carrier extraction in treated films compared to pristine directly measured by CELIV current transients. Improvement in the carrier mobility can be seen from the shift in the position of extraction maximum, while experimental conditions (film thicknesses and applied voltages) were kept similar. Thermal annealing of films fabricated with processing additive results in no change in performance. Reprinted with permission from (Pivrikas et al., 2008). Copyright 2008, with permission from Elsevier.

The dark current in the region of negative applied voltage (the reverse bias, positive voltage on Al, negative on ITO), is similar in all cells, showing that current injection is contact limited. A significant rectification ratio is observed for all types of studied cells. The dark leakage current in reverse bias is rather high, but similar for all cells.

Due to the different nanomorphologies of the interpenetrating network, the dark conductivity is expected to increase in the cells with higher conversion efficiency, because of improved conductivity of the films (assuming the injection is not limited by the contact). The dark injection current in forward bias is observed to be significantly higher in treated cells. In Fig. 17 (b) the dark injection current in forward bias is plotted in log-log scale for all devices. Faster charge carrier mobilities in all cells were estimated from these dependences using the Mott-Gurney Law. As can be directly seen from the magnitude of injection current, the highest mobility was observed in the films with chemical additives, confirming the beneficial effect of chemical additives for charge transport in bulk-heterojunction solar cells. From CELIV measurements, shown in Fig. 17 (c) it was demonstrated that charge carrier mobility is mainly reponsible for improvements in OPV performance.

However, the charge carrier recombination processes in operating devices has yet to be clarified. It was shown that the typically expected Langevin bimolecular charge carrier recombination can be avoided in highly efficiency P3HT and PCBM blends.(Pivrikas et al., 2005) Non-Langevin carrier recombination was shown to be crucially important in low mobility organic photovoltaic devices, since the requirement for the slower carrier mobility can be reduced without recombination losses. This implies that close to unity Internal quantum efficiency can be reached in low bandgap organic materials with very low carrier mobility if reduced bimolecular recombination (non-Langevin) is present in the device.

6. Conclusions

The film nanomorphology of bulk heterojunction solar cells determines the power conversion efficiency through photophysical properties such as light absorption, exciton dissociation, charge transport and recombination. The nano-morphology can be controlled by a variety of different methods. Thermal annealing of fabricated solar cells can be successfully substituted with slow drying of the solvent or chemical additives. These methods induce the phase separation between the donor and acceptor in the bulk-heterojunction, which results in red-shifted light absorption, improved exciton dissociation, faster charge carrier transport, and reduced recombination. Segregated donor-enriched and/or acceptor-enriched phases can be formed resulting in an interpenetrating bicontinuous network with the domain sizes comparable to the exciton diffusion length. Interconnected pathways for electromn and hole transport to the electrodes are required. This structure is essential for the photovoltaic performance of polymer-based solar cells. Therefore, reproducible, low cost nano-structure control is crucially important for fabrication of high efficiency OPV suitable for commercialization. In order to be able to control and predict the film nano-morphology of novel materials, an understanding of the material parameters governing the phase separation is required.

7. Acknowledgements

The author would like to Dr. Paul Schwenn for helpful discussions during manuscript preparation.

8. References

Alargova, R., Deguchi, S. & Tsujii, K. (2001). Stable colloidal dispersions of fullerenes in polar organic solvents, *Journal of the American Chemical Society* 123(43): 10460–10467.

Baranovski, S. (2006). *Charge transport in disordered solids with applications in electronics*, John Wiley & Sons Inc.

Bassler, H. (1993). Charge transport in disordered organic photoconductors a monte carlo simulation study, *physica status solidi (b)* 175(1): 15–56.

Bechara, R., Leclerc, N., Lévêque, P., Richard, F., Heiser, T. & Hadziioannou, G. (2008). Efficiency enhancement of polymer photovoltaic devices using thieno-thiophene based copolymers as nucleating agents for polythiophene crystallization, *Applied Physics Letters* 93: 013306.

Brabec, C. (2004). Organic photovoltaics: technology and market, *Solar energy materials and solar cells* 83(2-3): 273–292.

Brabec, C., Sariciftci, N., Hummelen, J. et al. (2001). Plastic solar cells, *Advanced Functional Materials* 11(1): 15–26.

Campbell, A., Hodgkiss, J., Westenhoff, S., Howard, I., Marsh, R., McNeill, C., Friend, R. & Greenham, N. (2008). Low-temperature control of nanoscale morphology for high performance polymer photovoltaics, *Nano letters* 8(11): 3942–3947.

Chen, F., Tseng, H. & Ko, C. (2008). Solvent mixtures for improving device efficiency of polymer photovoltaic devices, *Applied Physics Letters* 92: 103316.

Chen, F., Xu, Q. & Yang, Y. (2004). Enhanced efficiency of plastic photovoltaic devices by blending with ionic solid electrolytes, *Applied physics letters* 84: 3181.

Chen, H., Yang, H., Yang, G., Sista, S., Zadoyan, R., Li, G. & Yang, Y. (2009). Fast-grown interpenetrating network in poly (3-hexylthiophene): Methanofullerenes solar cells processed with additive, *The Journal of Physical Chemistry C* 113(18): 7946–7953.

Chen, L., Hong, Z., Li, G. & Yang, Y. (2009). Recent progress in polymer solar cells: manipulation of polymer: fullerene morphology and the formation of efficient inverted polymer solar cells, *Advanced Materials* 21(14-15): 1434–1449.

Cho, S., Lee, J., Moon, J., Yuen, J., Lee, K. & Heeger, A. (2008). Bulk heterojunction bipolar field-effect transistors processed with alkane dithiol, *Organic Electronics* 9(6): 1107–1111.

Chu, C., Yang, H., Hou, W., Huang, J., Li, G. & Yang, Y. (2008). Control of the nanoscale crystallinity and phase separation in polymer solar cells, *Applied Physics Letters* 92: 103306.

Coates, N., Hwang, I., Peet, J., Bazan, G., Moses, D. & Heeger, A. (2008). 1, 8-octanedithiol as a processing additive for bulk heterojunction materials: Enhanced photoconductive response, *Applied Physics Letters* 93: 072105.

Cohen, E. & Gutoff, E. (1992). *Modern coating and drying technology*, VCH.

Cox, P., Betts, R., Jones, C., Spall, S. & Totterdell, I. (2000). Acceleration of global warming due to carbon-cycle feedbacks in a coupled climate model, *Nature* 408(6809): 184–187.

Deibel, C. & Dyakonov, V. (2010). Polymer–fullerene bulk heterojunction solar cells, *Reports on Progress in Physics* 73: 096401.

Dennler, G., Scharber, M. & Brabec, C. (2009). Polymer-fullerene bulk-heterojunction solar cells, *Advanced Materials* 21(13): 1323–1338.

Erb, T., Zhokhavets, U., Gobsch, G., Raleva, S., ST
"uhN, B., Schilinsky, P., Waldauf, C. & Brabec, C. (2005). Correlation between structural and optical properties of composite polymer/fullerene films for organic solar cells, *Advanced Functional Materials* 15(7): 1193–1196.

Forrest, S. (2005a). The limits to organic photovoltaic cell efficiency, *MRS bulletin* 30(01): 28–32.

Forrest, S. (2005b). The path to ubiquitous and low-cost organic electronic appliances on plastic, *Gravitational, electric, and magnetic forces: an anthology of current thought* p. 120.

Glaser, P. (1968). Power from the sun: Its future, *Science* 162(3856): 857.

Gunes, S., Neugebauer, H. & Sariciftci, N. (2007). Conjugated polymer-based organic solar cells, *Chemical reviews* 107(4): 1324–1338.

Halls, J., Walsh, C., Greenham, N., Marseglia, E., Friend, R., Moratti, S. & Holmes, A. (1995). Efficient photodiodes from interpenetrating polymer networks.

Hecht, K. (1932). Zum mechanismus des lichtelektrischen primärstromes in isoliereden kristallen, *Z. Phys* 77: 235.

Hoppe, H., Glatzel, T., Niggemann, M., Schwinger, W., Schaeffler, F., Hinsch, A., Lux-Steiner, M. & Sariciftci, N. (2006). Efficiency limiting morphological factors of mdmo-ppv: Pcbm plastic solar cells, *Thin solid films* 511: 587–592.

Hoppea, H. & Sariciftci, N. (2004). Organic solar cells: An overview, *J. Mater. Res* 19(7): 1925.

Huang, J., Chien, F., Chen, P., Ho, K. & Chu, C. (2010). Monitoring the 3d nanostructures of bulk heterojunction polymer solar cells using confocal lifetime imaging, *Analytical chemistry* 82(5): 1669–1673.

Hwang, I., Cho, S., Kim, J., Lee, K., Coates, N., Moses, D. & Heeger, A. (2008). Carrier generation and transport in bulk heterojunction films processed with 1, 8-octanedithiol as a processing additive, *Journal of Applied Physics* 104(3): 033706–033706.

Ihn, K., Moulton, J. & Smith, P. (1993). Whiskers of poly (3-alkylthiophene) s, *Journal of Polymer Science Part B: Polymer Physics* 31(6): 735–742.

Jorgensen, M., Norrman, K. & Krebs, F. (2008). Stability/degradation of polymer solar cells, *Solar Energy Materials and Solar Cells* 92(7): 686–714.

Juska, G., Genevičius, K.,

"Osterbacka, R., Arlauskas, K., Kreouzis, T., Bradley, D. & Stubb, H. (2003). Initial transport of photogenerated charge carriers in π-conjugated polymers, *Physical Review B* 67(8): 081201.

Kim, K., Liu, J. & Carroll, D. (2006). Thermal diffusion processes in bulk heterojunction formation for poly-3-hexylthiophene/c60 single heterojunction photovoltaics, *Applied physics letters* 88(18): 181911–181911.

Kim, Y., Choulis, S., Nelson, J., Bradley, D., Cook, S. & Durrant, J. (2005). Device annealing effect in organic solar cells with blends of regioregular poly (3-hexylthiophene) and soluble fullerene, *Applied Physics Letters* 86(6): 063502–063502.

Kim, Y., Cook, S., Choulis, S., Nelson, J., Durrant, J. & Bradley, D. (2005). Effect of electron-transport polymer addition to polymer/fullerene blend solar cells, *Synthetic metals* 152(1-3): 105–108.

Kim, Y., Cook, S., Tuladhar, S., Choulis, S., Nelson, J., Durrant, J., Bradley, D., Giles, M., McCulloch, I., Ha, C. et al. (2006). A strong regioregularity effect in self-organizing conjugated polymer films and high-efficiency polythiophene: fullerene solar cells, *nature materials* 5(3): 197–203.

Krebs, F. (2009). Fabrication and processing of polymer solar cells: a review of printing and coating techniques, *Solar Energy Materials and Solar Cells* 93(4): 394–412.

Lee, J., Ma, W., Brabec, C., Yuen, J., Moon, J., Kim, J., Lee, K., Bazan, G. & Heeger, A. (2008). Processing additives for improved efficiency from bulk heterojunction solar cells, *Journal of the American Chemical Society* 130(11): 3619–3623.

Li, G., Shrotriya, V., Huang, J., Yao, Y., Moriarty, T., Emery, K. & Yang, Y. (2005). High-efficiency solution processable polymer photovoltaic cells by self-organization of polymer blends, *Nature Materials* 4(11): 864–868.

Li, G., Shrotriya, V., Yao, Y., Huang, J. & Yang, Y. (2007). Manipulating regioregular poly (3-hexylthiophene):[6, 6]-phenyl-c61-butyric acid methyl ester blends-route towards high efficiency polymer solar cells, *Journal of Materials Chemistry* 17(30): 3126–3140.

Li, G., Yao, Y., Yang, H., Shrotriya, V., Yang, G. & Yang, Y. (2007). Şsolvent annealingŤ effect in polymer solar cells based on poly (3-hexylthiophene) and methanofullerenes, *Advanced Functional Materials* 17(10): 1636–1644.

Li, L., Lu, G. & Yang, X. (2008). Improving performance of polymer photovoltaic devices using an annealing-free approach via construction of ordered aggregates in solution, *J. Mater. Chem.* 18(17): 1984–1990.

Luque, A. & Hegedus, S. (2003). *Handbook of photovoltaic science and engineering*, John Wiley & Sons Inc.

Ma, W., Yang, C., Gong, X., Lee, K. & Heeger, A. (2005). Thermally stable, efficient polymer solar cells with nanoscale control of the interpenetrating network morphology, *Advanced Functional Materials* 15(10): 1617–1622.

Mayer, A., Toney, M., Scully, S., Rivnay, J., Brabec, C., Scharber, M., Koppe, M., Heeney, M., McCulloch, I. & McGehee, M. (2009). Bimolecular crystals of fullerenes in conjugated polymers and the implications of molecular mixing for solar cells, *Adv. Funct. Mater* 19(8): 1173–1179.

Mihailetchi, V., Xie, H., de Boer, B., Popescu, L., Hummelen, J., Blom, P. & Koster, L. (2006). Origin of the enhanced performance in poly (3-hexylthiophene):[6, 6]-phenyl c-butyric acid methyl ester solar cells upon slow drying of the active layer, *Applied physics letters* 89: 012107.

Moule, A. & Meerholz, K. (2008). Controlling morphology in polymer–fullerene mixtures, *Advanced Materials* 20(2): 240–245.

Moulee, A., Tsami, A., B "unnagel, T., Forster, M., Kronenberg, N., Scharber, M., Koppe, M., Morana, M., Brabec, C., Meerholz, K. et al. (2008). Two novel cyclopentadithiophene-based alternating copolymers as potential donor components for high-efficiency bulk-heterojunction-type solar cells, *Chemistry of Materials* 20(12): 4045–4050.

Nayak, P., Bisquert, J. & Cahen, D. (2011). Assessing possibilities and limits for solar cells, *Advanced Materials* .

Nelson, J. (2003). *The physics of solar cells*, Imperial College Press London.

Norrman, K., Ghanbari-Siahkali, A. & Larsen, N. (2005). 6 studies of spin-coated polymer films, *Annu. Rep. Prog. Chem., Sect. C: Phys. Chem.* 101: 174–201.

Oesterbacka, R., Pivrikas, A., Juska, G., Poskus, A., Aarnio, H., Sliauzys, G., Genevicius, K., Arlauskas, K. & Sariciftci, N. (2010). Effect of 2-d delocalization on charge transport and recombination in bulk-heterojunction solar cells, *IEEE Journal of Selected Topics in Quantum Electronics* 16(6): 1738–1745.

Osterbacka, R., Geneviius, K., Pivrikas, A., Juka, G., Arlauskas, K., Kreouzis, T., Bradley, D. & Stubb, H. (2003). Quantum efficiency and initial transport of photogenerated charge carriers in [pi]-conjugated polymers, *Synthetic metals* 139(3): 811–813.

Ouyang, J. & Xia, Y. (2009). High-performance polymer photovoltaic cells with thick p3ht: Pcbm films prepared by a quick drying process, *Solar Energy Materials and Solar Cells* 93(9): 1592–1597.

Padinger, F., Rittberger, R. & Sariciftci, N. (2003). Effects of postproduction treatment on plastic solar cells, *Advanced Functional Materials* 13(1): 85–88.

Peet, J., Brocker, E., Xu, Y. & Bazan, G. (2008). Controlled β-phase formation in poly (9, 9-di-n-octylfluorene) by processing with alkyl additives, *Advanced Materials* 20(10): 1882–1885.

Peet, J., Kim, J., Coates, N., Ma, W., Moses, D., Heeger, A. & Bazan, G. (2007). Efficiency enhancement in low-bandgap polymer solar cells by processing with alkane dithiols, *Nature Materials* 6(7): 497–500.

Peet, J., Senatore, M., Heeger, A. & Bazan, G. (2009). The role of processing in the fabrication and optimization of plastic solar cells, *Advanced Materials* 21(14-15): 1521–1527.

Peet, J., Soci, C., Coffin, R., Nguyen, T., Mikhailovsky, A., Moses, D. & Bazan, G. (2006). Method for increasing the photoconductive response in conjugated polymer/fullerene composites, *Applied physics letters* 89(25): 252105–252105.

Peumans, P., Uchida, S. & Forrest, S. (2003). Efficient bulk heterojunction photovoltaic cells using small-molecular-weight organic thin films, *Nature* 425(6954): 158–162.

Pivrikas, A., Juška, G., Mozer, A., Scharber, M., Arlauskas, K., Sariciftci, N., Stubb, H. & "Osterbacka, R. (2005). Bimolecular recombination coefficient as a sensitive testing parameter for low-mobility solar-cell materials, *Physical review letters* 94(17): 176806.

Pivrikas, A., Neugebauer, H. & Sariciftci, N. (2010a). Charge carrier lifetime and recombination in bulk heterojunction solar cells, *Selected Topics in Quantum Electronics, IEEE Journal of* 16(6): 1746–1758.

Pivrikas, A., Neugebauer, H. & Sariciftci, N. (2010b). Influence of processing additives to nano-morphology and efficiency of bulk-heterojunction solar cells: A comparative review, *Solar Energy* .

Pivrikas, A., Stadler, P., Neugebauer, H. & Sariciftci, N. (2008). Substituting the postproduction treatment for bulk-heterojunction solar cells using chemical additives, *Organic Electronics* 9(5): 775–782.

PIVRIKAS, A., ULLAH, M., SINGH, T., SIMBRUNNER, C., MATT, G., SITTER, H. & SARICIFTCI, N. (2011). Meyer-neldel rule for charge carrier transport in fullerene devices: A comparative study, *Organic electronics* 12(1): 161–168.

Pivrikas, A., Ullah, M., Sitter, H. & Sariciftci, N. (2011). Electric field dependent activation energy of electron transport in fullerene diodes and field effect transistors: Gillšs law, *Applied Physics Letters* 98: 092114.

Rughooputh, S., Hotta, S., Heeger, A. & Wudl, F. (1987). Chromism of soluble polythienylenes, *Journal of Polymer Science Part B: Polymer Physics* 25(5): 1071–1078.

Sariciftci, N. (2006). Morphology of polymer/fullerene bulk heterojunction solar cells, *Journal of Materials Chemistry* 16(1): 45–61.

Scharber, M., M
"uhlbacher, D., Koppe, M., Denk, P., Waldauf, C., Heeger, A. & Brabec, C. (2006). Design rules for donors in bulk-heterojunction solar cellsŮtowards 10% energy-conversion efficiency, *Advanced Materials* 18(6): 789–794.

Schwoerer, M. & Wolf, H. (2007). *Organic molecular solids*, Wiley Online Library.

Shaheen, S., Brabec, C., Sariciftci, N., Padinger, F., Fromherz, T. & Hummelen, J. (2001). 2.5% efficient organic plastic solar cells, *Applied Physics Letters* 78: 841.

Shockley, W. & Queisser, H. (1961). Detailed balance limit of efficiency of p-n junction solar cells, *Journal of Applied Physics* 32(3): 510–519.

Shrotriya, V., Li, G., Yao, Y., Moriarty, T., Emery, K. & Yang, Y. (2006). Accurate measurement and characterization of organic solar cells, *Advanced Functional Materials* 16(15): 2016–2023.

Sivula, K., Ball, Z., Watanabe, N. & Fréchet, J. (2006). Amphiphilic diblock copolymer compatibilizers and their effect on the morphology and performance of polythiophene: fullerene solar cells, *Advanced Materials* 18(2): 206–210.

Sun, S., Zhang, C., Ledbetter, A., Choi, S., Seo, K., Bonner, C., Drees, M. & Sariciftci, N. (2007). Photovoltaic enhancement of organic solar cells by a bridged donor-acceptor block copolymer approach, *Applied physics letters* 90(4): 043117–043117.

Treat, N., Brady, M., Smith, G., Toney, M., Kramer, E., Hawker, C. & Chabinyc, M. (2011). Interdiffusion of pcbm and p3ht reveals miscibility in a photovoltaically active blend, *Laser Physics Review* 1: 82–89.

Troshin, P., Hoppe, H., Renz, J., Egginger, M., Mayorova, J., Goryachev, A., Peregudov, A., Lyubovskaya, R., Gobsch, G., Sariciftci, N. et al. (2009). Material solubility-photovoltaic performance relationship in the design of novel fullerene derivatives for bulk heterojunction solar cells, *Adv. Funct. Mater* 19: 779–788.

Turner, J. (1999). A realizable renewable energy future, *Science* 285(5428): 687.

van Duren, J., Yang, X., Loos, J., Bulle-Lieuwma, C., Sieval, A., Hummelen, J. & Janssen, R. (2004). Relating the morphology of poly (p-phenylene vinylene)/methanofullerene blends to solar-cell performance, *Advanced Functional Materials* 14(5): 425–434.

Vanlaeke, P., Vanhoyland, G., Aernouts, T., Cheyns, D., Deibel, C., Manca, J., Heremans, P. & Poortmans, J. (2006). Polythiophene based bulk heterojunction solar cells: Morphology and its implications, *Thin Solid Films* 511: 358–361.

Wienk, M., Turbiez, M., Gilot, J. & Janssen, R. (2008). Narrow-bandgap diketo-pyrrolo-pyrrole polymer solar cells: The effect of processing on the performance, *Advanced Materials* 20(13): 2556–2560.

Wohrle, D. & Meissner, D. (1991). Organic solar cells, *Advanced Materials* 3(3): 129–138.

Xin, H., Kim, F. & Jenekhe, S. (2008). Highly efficient solar cells based on poly (3-butylthiophene) nanowires, *Journal of the American Chemical Society* 130(16): 5424–5425.

Xu, B. & Holdcroft, S. (1993). Molecular control of luminescence from poly (3-hexylthiophenes), *Macromolecules* 26(17): 4457–4460.

Yang, C., Qiao, J., Sun, Q., Jiang, K., Li, Y. & Li, Y. (2003). Improvement of the performance of polymer/c60 photovoltaic cells by small-molecule doping, *Synthetic metals* 137(1-3): 1521–1522.

Yang, X., Loos, J., Veenstra, S., Verhees, W., Wienk, M., Kroon, J., Michels, M. & Janssen, R. (2005). Nanoscale morphology of high-performance polymer solar cells, *Nano Letters* 5(4): 579–583.

Yao, Y., Hou, J., Xu, Z., Li, G. & Yang, Y. (2008). Effects of solvent mixtures on the nanoscale phase separation in polymer solar cells, *Advanced Functional Materials* 18(12): 1783–1789.

Yao, Y., Shi, C., Li, G., Shrotriya, V., Pei, Q. & Yang, Y. (2006). Effects of c70 derivative in low band gap polymer photovoltaic devices: Spectral complementation and morphology optimization, *Applied physics letters* 89(15): 153507–153507.

Yu, G., Gao, J., Hummelen, J., Wudl, F. & Heeger, A. (1995). Polymer photovoltaic cells: enhanced efficiencies via a network of internal donor-acceptor heterojunctions, *Science* 270(5243): 1789.

Zhang, C., Choi, S., Haliburton, J., Cleveland, T., Li, R., Sun, S., Ledbetter, A. & Bonner, C. (2006). Design, synthesis, and characterization of a-donor-bridge-acceptor-bridge-type block copolymer via alkoxy-and sulfone-derivatized poly (phenylenevinylenes), *Macromolecules* 39(13): 4317–4326.

Zhang, F., Jespersen, K., Björström, C., Svensson, M., Andersson, M., Sundstr"om, V., Magnusson, K., Moons, E., Yartsev, A. & Ingan"as, O. (2006). Influence of solvent mixing on the morphology and performance of solar cells based on polyfluorene copolymer/fullerene blends, *Advanced Functional Materials* 16(5): 667–674.

Zhokhavets, U., Erb, T., Hoppe, H., Gobsch, G. & Serdar Sariciftci, N. (2006). Effect of annealing of poly (3-hexylthiophene)/fullerene bulk heterojunction composites on structural and optical properties, *Thin Solid Films* 496(2): 679–682.

7

One-Step Physical Synthesis of Composite Thin Film

Seishi Abe
Research Institute for Electromagnetic Materials
Japan

1. Introduction

Quantum-dot solar cells have attracted much attention because of their potential to increase conversion efficiency of solar photo conversion up to almost 66% by utilizing hot photogenerated carriers to produce higher photovoltages or higer photocurrents (Nozik, 2002). Specifically, the optical-absorption edge of a semiconductor nanocrystal is often shifted due to the quantum-size effect. The optical band gap can then be tuned to the effective energy region for absorbing maximum intensity of the solar radiation spectrum (Landsberg et al., 1993; Kolodinski et al., 1993). Furthermore, quantum dots produce multiple electron-hole pairs per -photon through impact ionization, whereas bulk semiconductor produces one electron-hole pair per -photon.

Wide gap semiconductor sensitized by semiconductor nanocrystal is candidate material for such. The wide gap materials such as TiO_2 can only absorb the ultraviolet part of the solar radiation spectrum. Hence, the semiconductor nanocrystal supports absorbing visible (vis)- and near-infrared (NIR) -light. Up to now, various nanocrystalline materials [InP (Zaban et al., 1998), CdSe (Liu & Kamat, 1994), CdS (Weller, 1991; Zhu et al., 2010), PbS (Hoyer & Könenkamp, 1995), and Ge (Chatterjee et al., 2006)] have been investigated, for instance, as the sensitizer for TiO_2. Alternatively, a wide-gap semiconductor ZnO is also investigated, since the band gap and the energetic position of the valence band maximum and conduction band minimum of ZnO are very close to that of TiO_2 (Yang et al., 2009). Most of these composite materials were synthesized through chemical techniques, however, physical deposition, such as sputtering, is also useful. In addition, package synthesis of the composite thin film is favorable for low cost product of solar cell.

In this chapter, Ge/TiO_2 and $PbSe/ZnSe$ composite thin film are presented, and they were prepared through rf sputtering and hot wall deposition (HWD), with multiple resources for simultaneous deposition. The package synthesis needs the specific material design for each of the preparation techniques. In the rf sputtering, the substances for nanocrystal and matrix are appropriately selected according to the difference in heat of formation (Ohnuma et al., 1996). Specifically, Ge nanocrystals are thermodynamically stable in a TiO_2 matrix, since Ti is oxidized more prominently than Ge along the fact that the heat of formation of GeO_2 is greater than those of TiO_2 (Kubachevski & Alcock, 1979). Larger difference in the heat of formation [e.g., Ge/Al-O (Abe et al., 2008a)] can provide thermodynamically more stable nanocrystal. Hence, the crystalline Ge was homogeneously embedded in amorphous Al oxide matrix, and evaluated unevenness of the granule size was ranged from 2 to 3nm, according to high resolution electron microscopy (HREM). In the HWD, on the other hand, the substances for nanocrystal and matrix are also selected following thermodynamic insolubility. The HWD technique, which is a kind of thermal evaporation, causes unintentional increase of the substrate-temperature due to the thermal irradiation. Hence, simultaneous HWD evaporation from multiple resources often produces solid solution [e.g., $Pb_{1-x}Ca_xS$ (Abe & Masumoto, 1999)]. Hence, package synthesis of the composite thin film needs insolubility material system. The bulk PbSe-ZnSe system, for instance, is found to phase-separate at thermal equilibrium state (Oleinik et al., 1982). It is therefore expected that PbSe nanocrystals phase-separate from the ZnSe matrix in spite of the simultaneous evaporation from PbSe- and ZnSe-resource.

Accordingly, the two thermodynamic material-designs, heat of formation for rf sputtering and insolubility system for HWD, are employed here for package synthesis of composite thin film. This chapter focuses on one-step physical synthesis of Ge/TiO_2 composite thin films by rf sputtering and PbSe/ZnSe composite thin films by HWD, as candidate materials for quantum dot solar cell.

2. Ge/TiO₂ composite thin films

TiO_2 mainly has crystal structures of rutile, anatase and brookite. It is believed that the anatase structure is favorable for the matrix, since carrier mobility and photoconductivity in the anatase structure exceed those in the rutile structure (Tang et al., 1994). It is difficult to forecast how the crystal structure of the TiO_2 matrix will be formed in such composite films. In fact, Ge/TiO_2 films prepared by rf sputtering employing a mixture target of TiO_2 and Ge powder hitherto contained anatase- and rutile -structure almost equally (Chatterjee, 2008). Hence, it is investigated here that the composition of Ge/TiO_2 films is thoroughly varied for preparing the anatase structure of the TiO_2 matrix while retaining vis-NIR absorption of Ge quantum dots.

2.1 Anatase-dominant matrix in Ge/TiO₂ thin films prepared by rf sputtering

The present study employed a new method of preparing Ge/TiO_2 films using a composite target of a Ge chip set on a TiO_2 disk, and their composition has been thoroughly changed. Figure 2-1(a) depicts the X-ray diffraction (XRD) pattern of Ge/TiO_2 thin films as a function of Ge concentration. In this case, the additional oxygen ratio in argon is kept constant at 0%. Labels A through E indicate Ge concentrations of 0, 1.9, 6.8, 8.1, and 21at.% by adopting 0, 1, 2, 3, and 21 Ge chips. XRD patterns first exhibited an amorphous state in as-deposited films, and several diffraction peaks began to appear at 723 K when the post-annealing temperature was raised from 673 to 873K in 50K steps. These peaks were assigned to TiO_2, and the films were therefore crystallized at around 723K (not shown here). A single-phase rutile structure is observed at a Ge concentration of 0at.% in the figure, corresponding to simple preparation of pure TiO_2 thin film. In our preliminary experiment for preparing the TiO_2 thin films, a single-phase anatase structure was obtained for oxygen ratios exceeding 0.5% and the successive post-annealing treatment. An insufficient oxygen ratio thus seems to cause formation of the rutile structure. Next, with a slight addition of Ge in the pattern of B, distinct diffraction peaks of anatase structure begin to appear, and the (101) Bragg reflection is dominant. Further addition of Ge, as seen in patterns C and D, produces different behavior in orientation, increasing the peak intensity at (004) reflection of the anatase structure. Finally, dominant, broad peaks of Ge can be observed with excess Ge addition in pattern E. The average size of the Ge nanogranules is estimated to be about 6.6nm based on the full-width at half maximum of the XRD peak employing Scherrer's equation (Scherrer, 1918). According to the variation of Ge concentration, the anatase structure is favorably promoted in patterns B, C, and D.

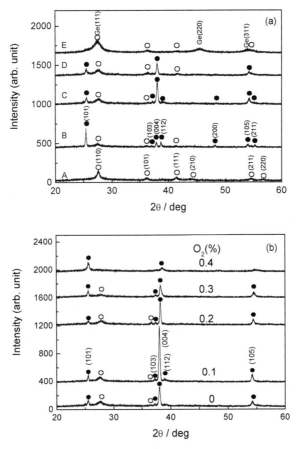

Fig. 2.1. (a) XRD patterns of Ge/TiO$_2$ composite films versus Ge concentrations.
(•) indicates anatase structure, and (○), rutile structure. (b) Same patterns versus
additional oxygen ratio in argon. (•) indicates anatase structure, and (○), rutile structure
(b) (after Abe et al., 2008b).

Figure 2-1(b) depicts the XRD pattern of Ge/TiO$_2$ thin films as a function of the additional
oxygen ratio in argon. In this case, the oxygen ratio is varied from 0 to 0.4%, and the
number of Ge chips is kept constant at 2. When the ratio is increased to 0.1%, the (004)
Bragg reflection becomes more prominent as seen in the figure. A further increase of the
oxygen ratio then indicates weakness. An anatase-dominant structure with strong
intensity at (004) reflection is thus observed at an oxygen ratio of 0.1%. We cannot observe
an XRD peak of Ge in the pattern within the precision of our experiment technique,
possibly due to the relatively low Ge concentration of 5.8at.%. This c-axis growth behavior
in an anatase-dominant structure seems to be unique even though the composite film is
deposited on a glass substrate. Thus, the crystal structure of TiO$_2$ matrix is found to be
changed with respect to the Ge number and the oxygen ratio as illustrated in Figs. 2-1(a)
and 2-1(b).

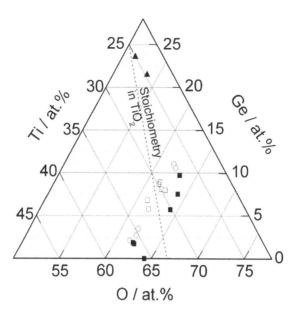

Fig. 2.2. Compositional plane of crystal structure of TiO₂ matrix in Ge/TiO₂ composite films. (○) indicates anatase structure, and (▲), rutile structure. (■) indicates coexistence of anatase and rutile structure. In particular, (□) indicates anatase-dominant structure with strong intensity at (004) reflection (after Abe et al., 2008b).

The relation between the analyzed composition of the films and the structure of TiO₂ matrix is summarized in Fig. 2-2 based on these results. The stoichiometric composition of TiO₂ is also plotted as a dotted line. The single phase of anatase structure (○) can be seen in the figure, but its visible absorption is quite weak. These films therefore do not achieve the present objective. A mixed phase containing anatase- and rutile -structure (■) appears in a wide range of Ge concentrations. In particular, an anatase-dominant structure with strong (004) reflection (□) is found at a Ge concentration of 6 to 9at.% near the stoichiometric composition of TiO₂. The optical absorption will be discussed using the following figure. The rutile structure (▲) is observed at a relatively high Ge concentration range. In these films, diffraction peaks of Ge nanogranules were observed at the same time [Fig. 2-1(a)]. Accordingly, the anatase-dominant structure with strong (004) reflection (□) is regarded to be the most optimized structure in the present study. As a further optimization, total gas pressure was varied from 2mTorr to 10mTorr in the optimized composition range. The (004) Bragg reflection was maximized at a gas pressure of 6mTorr; however, a slight amount of rutile structure still remained.

In the above sections, the structural optimization of the TiO₂ matrix in the Ge/TiO₂ composite films was focused. Next, we shall investigate the optical properties. Figure 2-3 depicts the typical optical absorption spectra of Ge/TiO₂ thin films thus optimized. For comparison, the spectrum of TiO₂ thin film is also presented in the figure. Ge has an indirect band-gap structure (Macfarlane et al., 1957), and the square root of absorbance is employed. As seen in the figure, the onset absorption can be confirmed at around 1.0eV in contrast to UV absorption of TiO₂ thin films due to its energy band gap of 3.2eV in the anatase structure. They can favorably cover the desirable energy region for high conversion efficiency (Loferski, 1956). Therefore, it should be pointed out that valuable characteristics of vis-NIR absorption and anatase-dominant structure of TiO_2 matrix are simultaneously retained in the Ge/TiO_2 composite thin films as a result of compositional optimization. Ge addition is first motivated to demonstrate the quantum size effect, then, it is worthy of note that its addition also effectively controls the crystal structure of the TiO_2 matrix. Consequently, a single phase of anatase structure cannot be obtained. However, extensive progress can be made in structural formation of the TiO_2 matrix as a result of exhaustive compositional investigation. Based on these results, Ge/TiO_2 thin films having an anatase-dominant structure of TiO_2 matrix and vis-NIR absorption should also be regarded as candidate materials for quantum dot solar cell.

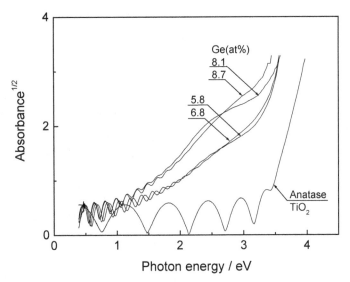

Fig. 2.3. Typical optical absorption spectra of Ge/TiO_2 composite films with anatase-dominant structure of TiO_2 matrix (after Abe et al., 2008b).

2.2 Solubility range and energy band gap of powder-synthesized $Ti_{1-x}Ge_xO_2$ solid solution

As a reason for the vis-NIR absorption, the quantum size effect probably appeared owing to the presence of Ge nanogranules. However, a ternary solid solution of $Ti_{1-x}Ge_xO_2$ is possibly formed as a matrix during the postannealing, and the solubility range of Ge and its energy band gap are hitherto unclear. Therefore, the reason for the vis-NIR absorption requires further investigation. To demonstrate whether the matrix exhibits the vis-NIR absorption, powder synthesis of a ternary $Ti_{1-x}Ge_xO_2$ solid solution is carried out. Specifically, the Ge/TiO_2 composite thin film contains multiple phases, and it is then difficult to focus on the matrix characteristics. In this section, $Ti_{1-x}Ge_xO_2$ solid solution is powder-synthesized, and the fundamental properties of solubility range of Ge and the energy band gap are investigated to clarify whether the ternary solid solution exhibits the vis-NIR absorption.

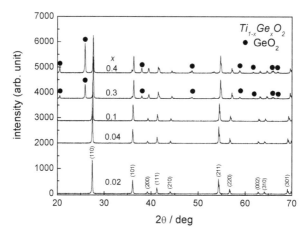

Fig. 2.4. Typical powder XRD patterns of $Ti_{1-x}Ge_xO_2$ solid solution with respect to x. Filled circle indicates GeO_2 (after Abe , 2009).

In a previous section, Ge nanogranules and TiO_2 matrix were thermally crystallized at an annealing temperature of 873K (Abe et al., 2008). Accordingly, a similar temperature of 923K was preliminary adopted to synthesize the $Ti_{1-x}Ge_xO_2$ solid solution. In this case, four samples (x = 0.05, 0.1, 0.2, and 0.3) were mixed and heat-treated for 20 days to achieve thermal equilibrium. However, a single phase of the $Ti_{1-x}Ge_xO_2$ solid solution could not be obtained, forming two phases of GeO_2 and anatase-structured TiO_2 according to the XRD pattern. For reference, there was a slight decrease in the lattice constant at x=0.05 estimated from the (004) reflection of anatase structure in comparison with those of the TiO_2 standard powder, and gradually increased with increasing x in the range exceeding 0.05. Thus, the solubility limit of Ge was found to be quite narrow (less than 0.05) at 923K. In addition, no energy shift of the optical absorption edge can be seen with respect to x. Therefore, an adequately high temperature of 1273K is alternatively adopted here in anticipation of a wide solubility range of Ge.

Figure 2-4 depicts typical powder XRD pattern of the $Ti_{1-x}Ge_xO_2$ solid solution. In the range below 0.1, all the XRD peaks can be assigned to rutile structure and shift toward greater angle as x increases owing to the difference in ionic radii between Ti and Ge (Shannon, 1976; Takahashi et al., 2006). In addition, an XRD peak of GeO_2 cannot be observed within the precision of the experimental technique. Such peak shift was also observed on the TiO_2-GeO_2 solid solution synthesized through sol-gel method within a Ge concentration range below 10 mol% (Kitiyanan et al., 2006) or 1.5 mol% (S. Chatterjee & A. Chatterjee, 2006). It is suggested that the present sample possibly forms a solid solution of $Ti_{1-x}Ge_xO_2$. The solubility range of Ge is therefore found to be enlarged as a result of elevating the temperature from 923 to 1273K. The standard powder of TiO_2 employed here has anatase structure, since the matrix of the Ge/TiO_2 composite thin films had anatase-dominant structure (Abe et al., 2008b). However, the product thus powder-synthesized resulted in rutile structure because of phase transition from anatase to rutile at 973K (The Mark Index, 1968). In contrast, the GeO_2 peaks, which are indicated by a filled circle, begin to appear in the range exceeding 0.3. Their peak positions seem to remain the same with respect to x, suggesting no solubility range of Ti in GeO_2 at 1273 K. The two phases of the $Ti_{1-x}Ge_xO_2$ and GeO_2 are therefore formed in such concentration range.

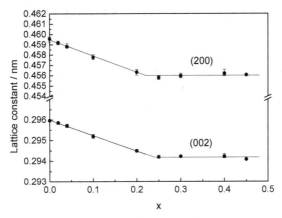

Fig. 2.5. Lattice constant of $Ti_{1-x}Ge_xO_2$ solid solution vs Ge concentration (after Abe , 2009).

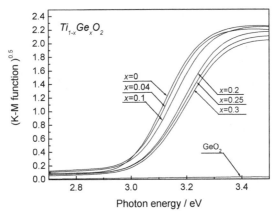

Fig. 2.6. Typical optical absorption spectra of $Ti_{1-x}Ge_xO_2$ solid solution vs Ge concentration (after Abe , 2009).

Next, the solubility limit of Ge in the $Ti_{1-x}Ge_xO_2$ is determined through the variation of the lattice constant. Figure 2-5 depicts the lattice constant of the $Ti_{1-x}Ge_xO_2$ solid solution as a function of x. Here, the lattice constant of the tetragonal system is estimated from the (200) and (002) reflections. Their peak intensities were found to be relatively weak (Fig. 2-4), but the peak position can be distinctly determined from Lorentzian fitting of the spectra, containing a measurement error of about 0.06 deg in 2θ as a result of four repetitive measurements. Accordingly, the lattice constant results in containing a maximum calculation error of about 0.0006 nm. In the preliminary experiment, a mass reduction during the heat treatment was found to be less than 0.1% in standard powders of TiO_2 and GeO_2, suggesting a small amount of sublimation. The nominal content of Ge is therefore employed here as a composition of the product. It is clearly seen in the figure that the lattice constant in both reflections is first decreased linearly in proportion to x, and becomes constant irrespective of x in the range exceeding 0.25. According to Vegard's law (Vegard, 1921), an on-setting composition x to deviate from the linearity is regarded as a solubility limit of Ge. It is therefore determined to be 0.23 ± 0.01 at 1273 K, having 0.22 at (200) and 0.24 at (002) reflection. Thus, the two-phase region consistently involves the fixed composition of $Ti_{0.77}Ge_{0.23}O_2$ and GeO_2. It can be clearly explained in terms of Gibb's phase rule. Specifically, the number of degrees of freedom is simply estimated to be 1, since the present ternary system has a component of 2. Thus, the state of the system is completely fixed at a given temperature of 1273K. Based on these results, the relation between the lattice constant L (nm) and x in the solubility range is expressed as follows: $L=0.4595-0.0162x$ at (200) reflection and $L=0.2960-0.0075x$ at (002) reflection.

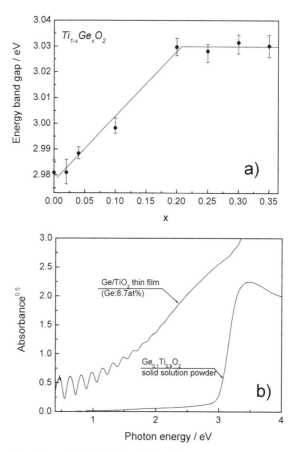

Fig. 2.7. (a) Energy band gap of $Ti_{1-x}Ge_xO_2$ solid solution at room temperature vs Ge concentration. (b) Absorption spectra of the Ge/TiO_2 thin film and a solid solution of the $Ti_{1-x}Ge_xO_2$ powder (after Abe , 2009).

In comparison, the lattice constant of the matrix in the Ge/TiO_2 composite thin films also decreased with increasing Ge concentration, ranging from 0.9495 to 0.9400nm estimated from the (004) reflection of anatase structure. In this case, the postannealing was performed at 873K for 60min. Such decreasing tendency of the lattice constant is the same as those of the $Ti_{1-x}Ge_xO_2$ powder despite the fact that the crystal structure is different in both samples, having a rutile structure for the powder and an anatase structure for the composite films. Therefore, the matrix of the composite film possibly formed a solid solution of $Ti_{1-x}Ge_xO_2$. Subsequently, optical absorption should be investigated regardless of whether the solid solution exhibits the vis-NIR absorption.

Figure 2-6 plots the optical absorption spectra of the powder-synthesized $Ti_{1-x}Ge_xO_2$ solid solution. These spectra are derived from the square root of Kubelka-Munk function (Kubelka & Munk, 1931) because of the indirect band gap structure of TiO_2 (Macfarlane et al., 1957). For comparison, the spectrum of GeO_2 is also shown. It is clearly seen that the GeO_2 is appreciably transparent in the measured range from 2.7 to 3.5 eV, whereas the optical absorption edge of the $Ti_{1-x}Ge_xO_2$ can be clearly observed at approximately 3 eV, and shift to the greater energy region as x increases. Therefore, the solid solution of $Ti_{1-x}Ge_xO_2$ is found to unexhibit the vis-NIR absorption.

Just for reference, the band gap can be estimated from a linear extrapolation to zero of the optical absorption edge, and is then summarized in Fig. 2-7(a). Error bars indicative of the possible variation in energy gap are used to plot the data. The energy band gap increases monotonically from 2.98 to 3.03 eV with respect to x, and becomes constant in the range exceeding 0.25. The energy shift is therefore achieved to be 0.05 eV at a solubility limit of 0.23. In fact, the Ge/TiO_2 composite thin films were compositionally optimized at a relatively low Ge concentration of 6 to 9at.% (Abe et al., 2008b), which indicates the total amount of Ge contained in both the matrix and the nanogranules. Hence, the energy shift in the matrix of the composite thin film is considered to be negligibly small (less than 0.01 eV). From these results, the matrix of the Ge/TiO_2 composite thin films possibly formed a solid solution of $Ti_{1-x}Ge_xO_2$ during the post annealing, but did not exhibit the vis-NIR absorption. Figure 2-7(b) depicts the optical absorption spectra of the Ge/TiO_2 thin film and a solid solution of the $Ti_{1-x}Ge_xO_2$ powder. Here, Ge concentration in the film was analyzed to be 8.7 at.%, and a similar concentration of $x=0.1$ in the powder was also presented for comparison. The absorption spectrum of the $Ti_{1-x}Ge_xO_2$ powder was obtained by means of the Kubelka-Munk function (Kubelka & Munk, 1931) through a diffused reflectance spectrum. The vis-NIR absorption was clearly observed in the Ge/TiO_2 film, while an optical absorption edge of the synthesized powder was observed at UV region. It was therefore concluded that the $Ti_{1-x}Ge_xO_2$ solid solution unexhibited such vis-NIR absorption.

2.3 Quantum size effect of Ge in TiO₂ matrix

In the previous sections, valuable characteristics of the vis-NIR absorption and the anatase-dominant structure of TiO_2 matrix were simultaneously retained in the Ge/TiO_2 composite thin films. However, it was unclear whether the vis-NIR absorption (Fig. 2-3) was due to the presence of Ge nanogranules, since an X-ray diffraction peak of Ge was not observed in the optimized composition range. In this section, we have investigated the presence of Ge nanogranules embedded in the anatase-dominant structure of TiO_2 thin films, and clarified the reason for the vis-NIR absorption.

Figure 2-8 depicts the size distribution of nanogranules in the Ge/TiO_2 composite thin films. These profiles were estimated from small angle X-ray spectroscopy (SAXS) analysis of Guinier fitting for an experimental result (the inset in Fig.2-8). In this case, Ge chips of 3 and the oxygen ratio of 0.3% was adopted during the deposition, and Ge concentration was analyzed to be 8.7at.%, and the film exhibited the vis-NIR absorption [Fig. 2-7(b)]. In the pinhole-collimated apparatus of SAXS measurement, X-ray was injected perpendicularly to the film surface, providing in-plane structural characteristic. As can be seen in the figure, the size profile distributed broadly, and mean radius of nanogranules was estimated to be 1.9 nm (3.8nm in diameter), ranging the radius from ~0.5 to 8nm. The SAXS analysis therefore strongly suggested that nanoscale material was embedded in the film, possibly attributing to Ge nanogranules or another phase of the $Ti_{1-x}Ge_xO_2$ matrix. Successively, the size of $Ti_{1-x}Ge_xO_2$ matrix was estimated by HREM.

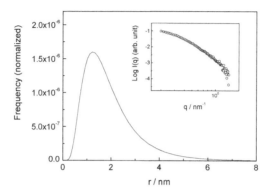

Fig. 2.8. Size distribution of nanogranules derived from the SAXS analysis. The inset depicts SAXS spectrum (after Abe et al., 2008c).

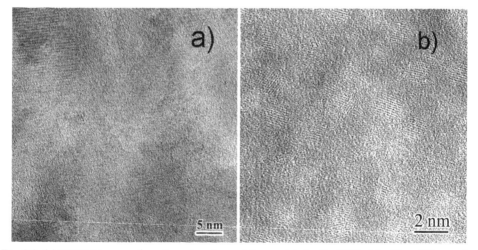

Fig. 2.9. (a) HREM image of anatase structure of the matrix in Ge/TiO_2 thin films (Abe et al., 2008c). (b) HREM image of Ge nanogranules embedded in the matrix (after Abe et al., 2008b).

Figure 2-9(a) presents the HREM image of the anatase-structured matrix at an oxygen ratio of 0.3%. Lattice image of the anatase structure was clearly observed, and size of their grains was estimated to be ~30nm. The size exceeded that of the nanogranules estimated by SAXS. Figure 2-9(b) presents an HREM image of Ge nanogranules embedded in Ge/TiO_2 composite film at a Ge concentration of 8.7at.%. In the figure, the slightly bright contrast region with spherical geometry corresponds to Ge nanogranule, and their lattice image can be clearly seen. The average size is estimated to be about 2nm. Furthermore, their average size is also estimated to be about 5nm by SAXS analysis. These estimated sizes are found to be close each other, and the Ge nanogranules are sufficiently small to create the quantum size effect because of the exciton Bohr radius of 24.3nm in Ge (Maeda et al., 1991). Therefore, the shift of optical absorption (Fig. 2-3) is reasonably due to the Ge nanogranules embedded in $Ti_{1-x}Ge_xO_2$ matrix.

3. PbSe/ZnSe composite thin films

Co-sputtering thus employed in the above section is useful for forming a composite thin film consisting of semiconductor nanocrystals embedded in a matrix because of its simple preparation process and consequent low cost. In the material design, based on the heat of formation, nanocrystal and matrix are clearly phase-separated in spite of the co-deposition from multiple sources (Abe et al., 2008b; Ohnuma et al., 1996). However, it is generally found that sputtering techniques often damage a film due to contamination of the fed gas and high-energy bombardment of the film surface. Thermal evaporation in a high-vacuum atmosphere seems to be better as a preparation technique from the point of view of film quality. In addition, the present study focuses on the insolubility of the material system, since simultaneous evaporation from multiple sources often provides a solid solution (Nill, et al., 1973; Holloway & Jesion, 1982; Abe & Masumoto, 1999).

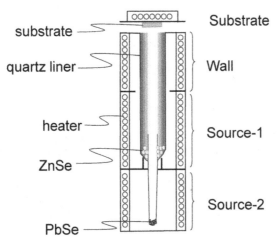

Fig. 3.1. HWD apparatus used in the study. It consists of four electric furnaces for substrate, wall, source-1, and source-2 (after Abe, 2011).

The PbSe-ZnSe system is a candidate for the composite. In the bulk thermal equilibrium state, the mutual solubility range is quite narrow, less than 1mol%, at temperatures below 1283K (Oleinik et al., 1982). In addition, a composite thin film of PbSe nanocrystal embedded in ZnSe matrix is capable of exhibiting the quantum size effect because of the relatively large exciton Bohr radius of 46nm in PbSe (Wise, 2000) and the relatively wide band gap of 2.67eV in ZnSe (Adachi & Taguchi, 1991). Hence, the optical gap of PbSe nanocrystals will probably be tuned to the maximum solar radiation spectrum. The dendritic PbSe nanostructure (Xue, 2009) and ZnSe nanobelt array (Liu, 2009), for instance, are hitherto investigated, but there is no report for one-step synthesis of PbSe/ZnSe composite thin film. Furthermore, an evaporation technique should be carefully selected, since the techniques involving a thermal non-equilibrium state, such as molecular beam epitaxy, increase the solubility limit (Koguchi et al., 1987). The use of HWD, which can provide an atmosphere near thermal equilibrium, is therefore indicated here (Lopez-Otero, 1978). Based on these considerations, one-step synthesis of a PbSe/ZnSe composite thin film was investigated by simultaneous HWD from multiple sources.

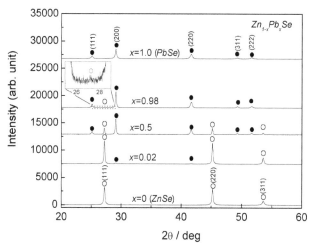

Fig. 3.2. XRD pattern of powder-synthesized $Zn_{1-x}Pb_xSe$ with respect to x. Dots indicate PbSe and circles indicate ZnSe (after Abe, 2011).

A PbSe/ZnSe composite thin film was prepared by the HWD method. Figure 3-1 is a schematic diagram of the HWD apparatus used. There were four electric furnaces in the apparatus, designated as substrate, wall, source-1, and source-2. Each temperature could be controlled independently. In the HWD method, deposition and re-evaporation are continuously repeated upon a film surface, resulting in achieving a state near thermal equilibrium (Lopez-Otero, 1973). PbSe and ZnSe were used as evaporation sources and were synthesized from elements of Pb, Zn, and Se with 6N purity. The PbSe and ZnSe sources were located at furnaces of source-2 and source-1 for simultaneous evaporation to a glass substrate (Corning #7059). Here, the temperatures were kept constant at 573K for the substrate, 773K for the wall, and 973K for source-1 (ZnSe). The source-2 (PbSe) temperature was varied from 763 to 833K to provide different PbSe concentrations.

The bulk PbSe-ZnSe phase diagram is now revealed at ZnSe concentrations below 45at.% (Pb-rich side) (Oleinik et al., 1982), although the phase diagram of the Zn-rich side still remains unclear. Powder synthesis of a PbSe-ZnSe system was investigated prior to investigating the film preparation. Figure 3-2 depicts the powder XRD pattern of the $Zn_{1-x}Pb_xSe$ system. In the powder synthesis, the bulk PbSe and ZnSe thus synthesized was used as starting materials. The desired composition of the system was prepared in an agate mortar and vacuum-sealed in a quartz tube for heat treatment at 1273K for 48h. Finally, the samples were successively water-quenched to maintain the solubility range at a synthesis temperature then crushed into powder for the following experiment setup. At $x=0$, all of the XRD peaks are assigned to the zinc-blend structure of ZnSe, with a lattice constant of 0.5669nm, estimated from the XRD peaks in a high-2θ range from $100°$ to $155°$, using the Nelson-Riley function (Nelson & Riley, 1945). The XRD peak of PbSe with an NaCl structure appears at Pb concentrations exceeding 0.02. The lattice constant of the ZnSe at $x=0.02$ is the same as at $x=0$, within the precision of the experiment technique. This result indicates that the solubility range of Pb in ZnSe is negligible. In contrast, the lattice constant of PbSe is estimated to be 0.6121nm at $x=1.0$ and 0.6117nm at $x=0.98$. A slight decrease in the lattice constant is seen in PbSe, due to the difference in ionic radii of Pb and Zn. Weak XRD peaks of ZnSe are also observed at $x=0.98$ as seen in the inset for easier viewing. This result indicates that the solubility range of Zn in PbSe is less than 0.02 at 1273K. The result is in good agreement with the previous result (Oleinik et al., 1982). The phase separation of the PbSe-ZnSe system is thus also seen on the Zn-rich side in the thermal-equilibrium state. The film preparation for PbSe/ZnSe composite is next investigated based on these results.

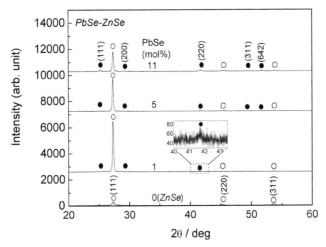

Fig. 3.3. XRD pattern of the PbSe/ZnSe composite thin films. Dots indicate PbSe and circles indicate ZnSe (after Abe, 2011).

The two sources were simultaneously evaporated to prepare a PbSe/ZnSe composite thin film. In the apparatus used, thermal radiation from the wall- and the source-furnace induced an unintentional increase of the substrate temperature up to 515K without use of the substrate-furnace. The deposition rate of the film was almost the same irrespective of the substrate temperature in the range from 515K to 593K. A homogeneous color is observed visually in these films. Above a substrate temperature of 593K, the deposition rate abruptly decreased with increasing temperature, since re-evaporation of PbSe from the film surface became dominant. The films visually exhibit an inhomogeneous yellowish and metallic color, probably caused by a significant reduction in the PbSe while the ZnSe remained, due to the relatively high vapor pressure of PbSe (Mills, 1974). The wall temperature also induced similar behavior. A substrate temperature of 573K and a wall temperature of 773K are therefore adopted throughout the present study.

Figure 3-3 depicts the XRD pattern for the PbSe/ZnSe composite thin films. The weak XRD peak of PbSe at 1mol% is enlarged in the inset for easier viewing. At a PbSe concentration of 0mol% (i.e., pure ZnSe), polycrystalline ZnSe with a zinc-blend structure is observed, with PbSe phase appearing at concentrations exceeding 1mol%. The solubility range of Pb in ZnSe is therefore found to be quite narrow, less than 1mol%, corresponding well to the bulk result (Fig. 3-2). The composite films thus deposited on a glass substrate exhibit a reasonably polycrystalline structure, but dominant (111) growth is seen in the ZnSe phase irrespective of x. At 1mol%, the lattice constant at the PbSe (220) peak is estimated to be 0.6118nm, close to that of the bulk result (Fig. 3-2). This result suggests that there is also a narrow solubility range on the Pb-rich side. The phase-separating PbSe-ZnSe system is therefore maintained not only in the bulk product, but also in the film thus obtained, despite the simultaneous evaporation from multiple sources. This result demonstrates that an atmosphere near thermal equilibrium was achieved in the HWD apparatus used.

Figure 3-4(a) presents a bright-field TEM image of the PbSe/ZnSe composite thin film containing 5mol%PbSe. Dark isolated grains with sizes of 25nm to 50nm are seen dispersed along the grain boundary of the bright area. Figures 3-4(b-e) present an scanning transmission electron microscopy (STEM) - energy dispersive spectroscopy (EDX) elemental mapping of the sample through X-ray detection of Zn K (red), Se K (blue), and Pb L (green). Similar morphology is also seen in the bright-field STEM image [Fig. 3-4(b)]. The dark grains indicate the absence of elemental Zn [Fig. 3-4(c)] and the presence of Se and Pb [Figs. 3-4 (d and e)]. It is thus determined that the dark grains are nanocrystalline PbSe. The other region is widely covered with the elements Zn and Se [Figs. 3-4 (c and d)], reasonably assumed to compose ZnSe. It is therefore determined that isolated PbSe nanocrystals are dispersed in the ZnSe matrix. The nanocrystals are estimated to be sufficiently small to exhibit the quantum-size effect because of the exciton Bohr radius of 46nm in PbSe (Wise, 2000).

Figure 3-5 depicts optical absorption spectra for the PbSe/ZnSe composite thin films. For comparison, the spectrum of a pure ZnSe thin film is also presented in the figure. PbSe and ZnSe have direct band structures (Theis, 1977; Zemel et al., 1965) and the absorbance squared is employed here. At a 0mol%PbSe, the optical absorption edge of ZnSe is clearly observed at 2.7eV. Weak absorption then broadly appears at a PbSe concentration of 1mol% in the visible region, together with the optical absorption edge of ZnSe. Such multiple absorptions are also seen in the spectra at concentrations up to 7mol%, indicating the obvious phase separation of the PbSe-ZnSe system. The broad absorption edge shifts toward the lower-energy region as the PbSe content increases. In particular, onset absorption can be confirmed at approximately 1.0eV at 16mol%PbSe, favorably covering the desirable energy region for high conversion efficiency (Loferski, 1956). Therefore, it should be noted that the PbSe/ZnSe composite thin film exhibits the valuable characteristic of vis-NIR absorption.

However, it is unclear whether the shift of the optical absorption edge is due to the PbSe nanocrystals, since the mean grain size of the PbSe remains almost the same at 27nm irrespective of the PbSe content, according to the XRD result (Fig. 3-3) using Scherrer's equation (Scherrer, 1918). A PbSe-ZnSe solid-solution system cannot provide continuous change of the energy band gap because of the quite narrow solubility range (Fig. 3-3). In contrast, the PbSe nanocrystals are sufficiently smaller than the exciton Bohr radius of PbSe (Fig. 3-4). Therefore, this obvious shift is assumed to be responsible for the quantum-size effect of the PbSe nanocrystals embedded in the ZnSe matrix. The minimal appearance of infrared absorption at 16mol%PbSe strongly suggests that relatively large-scale PbSe grains are partially involved in the composite film, since the energy band gap of bulk PbSe is

0.27eV (Zemel et al., 1965). Another TEM image also indicates the presence of relatively large PbSe crystals of approximately 100nm, even with a small amount of 5mol%PbSe (not shown here). Hence, the mean grain size of the PbSe is bimodally distributed in the composite. These large-scale PbSe grains probably dominate the full width at half maximum value of the XRD peak, resulting in no obvious relation between the optical absorption shift and the PbSe grain size. The size control of the nanocrystalline PbSe is therefore insufficient in the present study. The substrate temperature thus adopted seems to assist in the aggregation of PbSe nanocrystals. However, a one-step synthesis of the composite package has the potential to lead to low-cost production of next-generation solar cells.

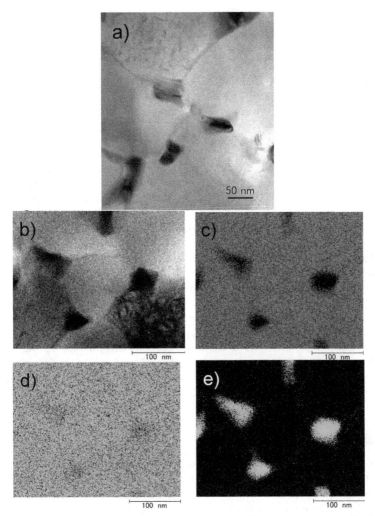

Fig. 3.4. Direct observation of PbSe/ZnSe composite thin film containing 5mol%PbSe. (a) Bright-field TEM image. (b) Bright-field image of STEM mode. (c) Elemental mapping of Zn (red), (d) Se (blue), and (e) Pb (green) (after Abe, 2011).

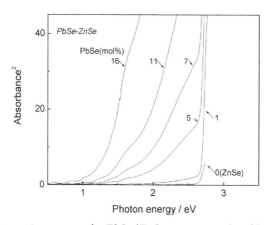

Fig. 3.5. Optical absorption spectra for PbSe/ZnSe composite thin films (after Abe, 2011).

4. Conclusion

This chapter has been focused on one-step physical synthesis of Ge/TiO$_2$ and PbSe/ZnSe composite thin film as candidate materials for quantum dot solar cell. It should be pointed out in the Ge/TiO$_2$ that the anatase-dominant structure appears in the restricted composition range as a result of optimization of Ge chip numbers and additional oxygen ratio in argon. Furthermore, their optical absorption edge is obviously shifted to vis-NIR region. The solubility range of Ge in the Ti$_{1-x}$Ge$_x$O$_2$ powder is estimated to be 0.23 ± 0.01 at 1273 K. In addition, their optical absorption edge is obviously shifted to the UV region as x increases. Thus, the Ti$_{1-x}$Ge$_x$O$_2$ solid solution does not exhibit the vis-NIR absorption. In contrast, SAXS and HREM results clearly indicated that the Ge nanogranules were embedded in the matrix. The size was sufficiently small to appear the quantum size effect. Thus, the both valuable characteristics are simultaneously retained in the Ge/TiO$_2$ composite films. In the PbSe/ZnSe, the solubility limit of Pb in ZnSe is quite narrow, less than 1mol% in the film form, indicating that an atmosphere near thermal equilibrium is achieved in the apparatus used. Elemental mapping indicates that isolated PbSe nanocrystals are dispersed in the ZnSe matrix. The optical absorption edge shifts toward the lower-photon-energy region as the PbSe content increases. In particular, onset absorption can be confirmed at approximately 1.0eV with 16mol%PbSe, favorably covering the desirable energy region for high conversion efficiency. The insolubility material system and the HWD technique enable a one-step synthesis of PbSe/ZnSe composite thin film.

5. Acknowledgment

The present work was supported in part by a Grant-in-Aid for Scientific Research from the Japan Society for the Promotion of Science (No.21360346). The author is grateful to Dr. M. Ohnuma [National Institute for Materials Science (NIMS), Tsukuba, Japan], Dr. D. H. Ping (NIMS), and Dr. S. Ohnuma [Research Institute for Electromagnetic Materials (RIEM), Sendai, Japan] for collaborating in this work. The author gratefully acknowledges the valuable comments and continuous encouragement of President T. Masumoto (RIEM). The author is also grateful to Mr. N. Hoshi and Mr. Y. Sato (RIEM) for assisting in the experiments.

6. References

Abe, S. & Masumoto, K. (1999). Compositional plane and properties of solid solution semiconductor $Pb_{1-x}Ca_xS_{1-y}Se_y$ for mid-infrared lasers, *Journal of Crystal Growth*, Vol.204, pp. 115-121.

Abe, S, Ohnuma, M, Ping, D. H., Ohnuma, S. (2008a). Single dominant distribution of Ge nanogranule embedded in Al-oxide thin film, *Journal of Aplied Physics*, Vol. 104, pp. 104305 1-3.

Abe, S, Ohnuma, M, Ping, D. H., Ohnuma, S (2008b). Anatase-Dominant Matrix in Ge/TiO_2 Thin Films Prepared by RF Sputtering Method , *Applied Physics Express*, Vol. 1, pp. 095001 1-3.

Abe, S, Ohnuma, M, Ping, D. H., Ohnuma, S (2008c). Preparation of Ge nanogranules embedded in Anatase-dominant TiO_2 thin films by RF sputtering, *Proceedings of 14th International Conference on Thin Films (ICTF 14)*, pp.101-104.

Abe, S (2009). Solubility Range and Energy Band Gap of Powder-Synthesized $Ti_{1-x}Ge_xO_2$ Solid Solution , *Japanese Journal of Applied Physics*, Vol. 48, pp. 081605 1-3

Abe, S (2011). One-step synthesis of PbSe/ZnSe composite thin film, *Nanoscale Research Letters* Vol. 6, pp.324 1-6.

Adachi, S. & Taguchi, T. (1991). Optical properties of $ZnSe$,: *Physical Review B*, Vol. 43, pp. 9569-9577.

Chatterjee, S. Goyal, A. & Shah, I. (2006). Inorganic nanocomposites for next generation photovoltaics, *Materials Letters*. Vol.60, pp. 3541-3543.

Chatterjee, S. & Chatterjee, A. (2008). Optoelectronic properties of Ge-doped TiO_2 nanoparticles. *Japanese Journal of Applied Physics*, Vol. 47, pp. 1136-1139.

Chatterjee, S. (2008). Titania-germanium nanocomposite as a photovoltaic material, *Solar Energy*, Vol.82, pp. 95-99.

Holloway, H. & Jesion, G. (1982). Lead strontium sulfide and lead calcium sulfide, two new alloy semiconductors, *Physical Review B*, Vol. 26, pp. 5617-5622.

Hoyer, P. & Könenkamp, R. (1995). Photoconduction in porus TiO_2 sensitized by PbS quantum dots, *Appied Physics Letters*, Vol. 66, pp. 349-351.

Kitiyanan, A. Kato, T. Suzuki, Y. & Yoshikawa, S. (2006). The use of binary TiO_2-GeO_2 oxide electrodes to enhanced efficiency of dye-sensitized solar cells, *Journal of Photochemistry and Photobiology A*, Vol. 179, pp. 130-134.

Koguchi, N. Kiyosawa, T.& Takahashi, S. (1987). Double hetero structure of $Pb_{1-x}Cd_xS_{1-y}Se_y$ lasers grown by molecular beam epitaxy, *Journal of Crystal Growth*, Vol. 81, pp. 400-404.

Kolodinski, S. Werner, J. H. Wittchen, T. & Queisser, H.J. (1993). Quantum efficiencies exceeding unity due to impact ionization in silicon solar cells, *Applied Physics Letters*, Vol. 63, pp. 2405-2407.

Kubachevski, O & Alcock, C. B. (1979). *Metallurgical Thermochemistry*, Pergamon.

Kubelka, P. & Munk, F. (1931). Ein Beitrag zur Optik der Far-banstriche, *Zeitschrift Technische Physik* Vol. 12, pp. 593-601.

Landsberg, P. T. Nussbaumer, H. & Willeke, G. (1993). Band–band impact ionization and solar cell efficiency , *Journal of Applied Physics*, Vol. 74, pp. 1451-1452.

Liu, D. & Kamat, P. V. (1993). Photoelectrochemical behavior of thin CdSe and coupled TiO_2/CdSe semiconductor films, *Journal of Physical Chemistry*, Vol. 97, pp. 10769-10773.

Liu, J. & Xue, D. (2009). Solution-based route to semiconductor film: Well-aligned ZnSe nanobelt arrays, *Thin Solid Films*, Vol. 517, pp. 4814-4817.

Loferski, J. J. (1956). Theoretical considerations covering the choice of the optinum semiconductor for photovoltaic solar energy conversion, *Journal of Applied Physics*, Vol. 27, pp.777-784.

Lopez-Otero, A. (1978). Hot wall epitaxy, *Thin Solid Films*, Vol. 49, pp. 3-57.

Macfarlane, G. G. McLean, T. P. Quarrington, J. E. & Roberts, V. (1957). Fine structure in the absorption-edge spectrum of Ge, *Physical Review* , Vol. 108, pp. 1377-1383.

Maeda, Y. Tsukamoto, N. Yazawa, Y. Kanemitsu, Y. & Masumoto, Y. (1991). Visible photoluminescence of Ge microcrystals embeddded inSiO_2, *Applied Physics Letters*, Vol. 59, pp. 3168-3170.

Mills, K. C. (1974). *Thermodynamic data for inorganic sulphide, selenides and Tellurides*. Butterworth.

Nelson, J. B. & Riley, D. P. (1945). An experimental investigation of extrapolation methods in the derivation of accurate unit-cell dimensions of crystals, *Proceedings of Physical Society* Vol. 57, pp. 160.

Nill, K. W. Sreauss, A. J. & Blum, F. A. (1973). Tunable cw Pb0.98Cd0.02S diode lasers emitting at 3.5 μm: Applications to ultrahigh-resolution spectroscopy, *Applied Physics Letters* Vol. 22, pp. 677-679.

Nozik, A. J. (2002). Quantum dot solar cells, *Physics E*, Vol. 14, pp.115-120.

Ohnuma, S. Fujimori, H. Mitani, S. & Masumoto, T. (1996). High-frequency magnetic properties in metal-nonmetal granular films, *Journal of Applied Physics*, Vol. 79, pp. 5130-5135.

Oleinik, G.S. Mizetskii, P.A. & Nizkova, A.I. (1982). Nature of the interaction between lead and zic chalcogenides, *Inorganic Materials*, Vol. 18, pp. 734-735.

Scherrer, P. (1918). Bestimmung der Größe und der inneren Struktur von Kolloidteilchen mittels Röntgenstrahlen, *Göttinger Nachrichten*, Vol. 2, pp. 98-100.

Shannon, R. D. (1976). Revised effective ionic radii and systematic studies of interatomic distances in halides and chalcogenides, *Acta Crystallography, Sect. A*, Vol. 32, pp. 751-767.

The Merck Index (Merck & Co, New Jersey, 1968) 8th ed., p. 1054.

Takahasi, Y. Kitamura, K. Iyi, N. & Inoue, S. (2006). Phase-stability and photoluminescence of BaTi(Si, Ge)$_3$O$_9$, *Journal of Ceramic Society of Japan*, Vol. 114, pp. 313-317.

Tang, H. Prasad, K. Sanjinès, R.P. Schmid, E. & Lévy F. (1994). Electrical and optical properties of TiO$_2$ anatase thin films, *Journal of Appied Physics*, Vol. 75, pp. 2042-2047.

Theis, D. (1977). Wavelength modulated reflectivity spectra of ZnSe and ZnS from 2.5 to 8 eV, *Physica Status Solidi* (B), Vol. 79, pp.125-130.

Vegard, L. (1921). Die Konstitution der Mischkristalle und die Raumerfüllung der Atome, *Z. Phys.* Vol. 5, pp. 17-26.

Weller, H.. (1991). Quantum sized semiconcuctor particles in solution in modified layers, *Berichte der Bunsengesellschaft Physical Chemistry*, Vol. 95, pp. 1361-1365.

Wise, F. W. (2000). Lead salts quantum dots: the limit of strong confinement, *Accounts of Chemical Research*, Vol. 33, pp. 773-780.

Xue, D. (2009). A template-free solution method based on solid-liquid interface reaction towards dendritic PbSe nanostructures, *Modern Physics Letters B*, Vol. 23, pp. 3817-3823.

Yang, W. Wan, F. Chen, S. Jiang, C. (2009). Hydrothermal growth and application of ZnO nanowire films with ZnO and TiO$_2$ buffer layers in dye-sensitized solar cells, *Nanoscale Research Letters*, Vol. 4, pp. 1486-1492.

Zaban, A. Micic, O. I. Gregg, B. A. & Nozik, A. J., (1998). Photosensitization of nanoporus TiO$_2$ electrodes with InP quantum dots, *Langmuir*, Vol. 14, pp. 3153-3156.

Zemel, J. N. Jensen, J. D. & Schoolar, R. B. (1965). Electrical and optical properties of epitaxial films of PbS, PbSe, PbTe, and SnTe, *Physical Review*, Vol. 140, pp. A330-A342.

Zhu, G. Su, F. Lv, T. Pan, L. Sun, Z. (2010). Au nanoparticles as interfacial layer for CdS quantum dot-sensitized solar cells, *Nanoscale Research Letters*, Vol. 5, pp. 1749-1754.

8

Bioelectrochemical Fixation of Carbon Dioxide with Electric Energy Generated by Solar Cell

Doo Hyun Park[1], Bo Young Jeon[1] and Il Lae Jung[2]
[1]Department of Biological Engineering, Seokyeong University, Seoul
[2]Department of Radiation Biology, Environmental Radiation Research Group,
Korea Atomic Energy Research Institute, Daejeon,
Korea

1. Introduction

Atmospheric carbon dioxide has been increased and was reached approximately to 390 mg/L at December 2010 (Tans, 2011). Rising trend of carbon dioxide in past and present time may be an indicator capable of estimating the concentration of atmospheric carbon dioxide in the future. Cause for increase of atmospheric carbon dioxide was already investigated and became general knowledge for the civilized peoples who are watching TV, listening to radio, and reading newspapers. Anybody of the civilized peoples can anticipate that the atmospheric carbon dioxide is increased continuously until unknowable time in the future but not in the near future. Carbon dioxide is believed to be a major factor affecting global climate variation because increase of atmospheric carbon dioxide is proportional to variation trend of global average temperature (Cox et al., 2000). Atmospheric carbon dioxide is generated naturally from the eruption of volcano (Gerlach et al., 2002; Williams et al., 1992), decay of organic matters, respiration of animals, and cellular respiration of microorganisms (Raich and Schlesinger, 2002; Van Veen et al., 1991); meanwhile, artificially from combustion of fossil fuels, combustion of organic matters, and cement making-process (Worrell et al., 2001). Theoretically, the natural atmospheric carbon dioxide generated biologically from the decay of organic matter and the respirations of organisms has to be fixed biologically by land plants, aquatic plants, and photosynthetic microorganisms, by which cycle of atmospheric carbon dioxide may be nearly balanced (Grulke et al., 1990). All of the human-emitted carbon dioxide except the naturally balanced one may be incorporated newly into the pool of atmospheric greenhouse gases that are methane, water vapor, fluorocarbons, nitrous oxide, and carbon dioxide (Lashof and Ahuja, 1990). The airborne fraction of carbon dioxide that is the ratio of the increase in atmospheric carbon dioxide to the emitted carbon dioxide variation was typically about 45% over 5 years period (Keeling et al., 1995). Canadell at al (2007) reported that about 57% of human-emitted carbon dioxide was removed by the biosphere and oceans. These reports indicate that the airborne fraction of carbon dioxide is at least 43-45%, which may be the balance emitted by human activity.

The land plants are the largest natural carbon dioxide sinker, which have been decreased globally by deforestation (Cramer et al., 2004). Especially, tropical and rainforests are being cut down for different purpose and by different reason and some of the forest are being burned for slash and burn farming. The atmospheric carbon dioxide and other greenhouse gases are increased in proportion to the deforestation (McKane et al. 1995). Deforestation causes part of the released carbon dioxide to be accumulated in the atmosphere and the global carbon cycle to be changed (Robertson and Tiejei, 1988). The releasing carbon dioxide and changing carbon cycle increase the greenhouse effect and may raise global temperature. The greenhouse effect is generated naturally by the infrared radiation, which is generated from incoming solar radiation, absorbed into atmospheric greenhouse gases and re-radiated in all direction (Held and Soden). The gases contributing to the greenhouse effect on Earth are water vapor (36-70%), carbon dioxide (9-26%), methane (4-9%), and ozone (3-7%) (Kiehl et al., 1977). Especially, water vapor can amplify the warming effect of other greenhouse gases, such that the warming brought about by increased carbon dioxide allows more water vapor to enter the atmosphere (Hansen, 2008). The greenhouse effect can be strengthened by human activity and enhanced by the synergetic effect of water vapor and carbon dioxide because the elevated carbon dioxide levels contribute to additional absorption and emission of thermal infrared in the atmosphere (Shine et al., 1999). The major non-gas contributor to the Earth's greenhouse effect, cloud (water vapor), also absorb and emit infrared radiation and thus have an effect on net warming of the atmosphere (Kiehl et al., 1997). Elevation of carbon dioxide is a cause for greenhouse effect, by which abnormal climate, desertification, and extinction of animals and plants may be induced (Stork, 1997). However, carbon dioxide is difficult to be controlled in the industry-based society that depends completely upon fossil fuel. If the elevation of carbon dioxide was unstoppable or necessary evil, the technique to convert biologically the atmospheric carbon dioxide to stable polymer in the condition without using fossil fuel must be developed. All of the land and aquatic plants convert mainly carbon dioxide to biomolecule in coupling with oxygen generation; however, a total of 16.5% of the forest (230,000 square miles) was affected by deforestation due to the increase of fragmented forests, cleared forests, and boundary areas between the fragmented forests (Skole et al., 1998). Decline of plants may be a cause to activate generation of the radiant heat because the visible radiation of solar energy absorbed for photosynthesis can be converted to additional radiant heat.

Solar cell is the useful equipment capable of physically absorbing solar radiation and converting the solar energy to electric energy (O'Regan and Grätzel, 1991). The radiant heat generated from the solar energy may be decreased in proportion to the electric energy produced by the solar cells. Electrochemical redox reaction can be generated from electric energy by using a specially designed bioreactor equipped with the anode and cathode separated with membrane, which is an electrochemical bioreactor. The electric energy generated from the solar energy can be converted to biochemical reducing power through the electrochemical redox mediator. The biochemical reducing power (NADH or NADPH) is the driving force to generate biochemical energy, ATP. The biochemical reducing power and ATP are essential elements that activate all biochemical reactions for biosynthesis of cell structure and production of metabolites.

2. Electrochemical redox mediator

The electrochemical reduction reaction generated in cathode can't catalyze reduction of NAD^+ or $NADP^+$ both *in vitro* and *in vivo* without electron mediator. Various ion radicals that are methyl viologen, benzyl viologen, hydroquinone, tetracyanoquinodimethane, and neutral red (NR) have been used as electron mediator to induce electrochemical redox reaction between electrode and electron carriers that are NAD+, FAD, and cytochrome C (Pollack et al., 1996; Park et al., 1997; Wang and Du, 2002; Kang et al., 2007). In order to *in vivo* drive and maintain bacterial metabolism with electrochemical reducing power as a sole energy source, only NAD+ or NADP+ is required to be reduced by coupling redox reaction between electron mediator and biochemical electron carrier (Park and Zeikus, 1999; 2000). NR can catalyze the electrochemical reduction reaction of NAD+ both *in vivo* and *in vitro* but no electron mediator except the NR can. NR is a water-soluble structure composed of phenazine ring with amine, dimethyl amine, methyl, and hydrogen group as shown in Fig 1. The dimethyl amine group is redox center for electron-accepting and donating in coupling with phenazine ring; meanwhile, the amine, methyl, and hydrogen are structural group. Redox potential of NR is -0.325 volt (vs. NHE), which is 0.05 volt lower than NAD+. The electrochemical redox reaction of NR can be coupled to biochemical redox reaction as follows:

$$[NR_{ox} + 2e^- + 1H^+ \rightarrow NR_{red}; NR_{red} + NAD^+ \rightarrow NR_{ox} + NADH]$$

NAD+ can be reduced in coupling with biochemical redox reaction as follows:

$$[NAD^+ + 2e^- + 2H^+ \rightarrow NADH + H^+]$$

Commonly, NR_{ox} and NAD+ are reduced to NR_{red} and NADH, respectively by accepting two electrons and one proton.

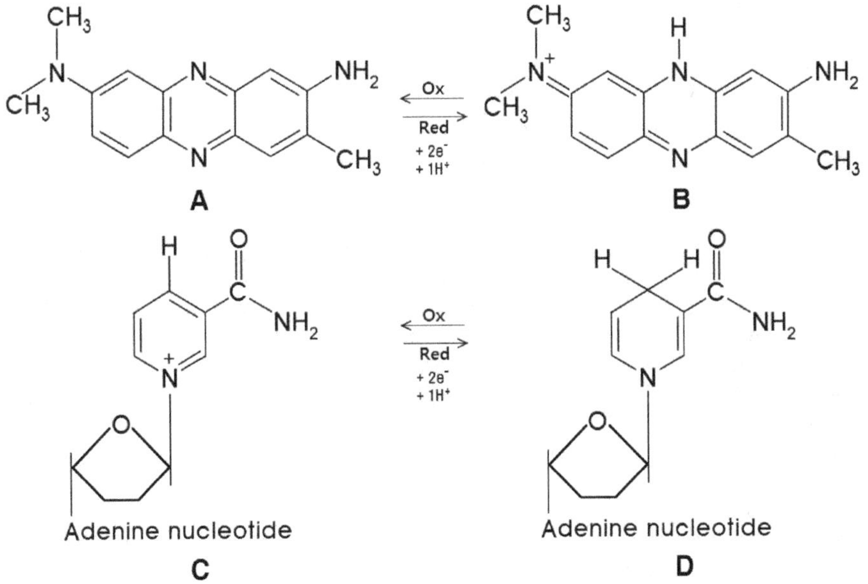

Fig. 1. Molecular structure of neutral red, which can be electrochemically oxidized (A) or reduced (B). The reduced neutral red can catalyze reduction reaction of NAD+ (C) to NADH (D) without enzyme catalysis. Ox and Red indicate oxidation and reduction, respectively.

Theoretically, the water-soluble NR may be reduced at the moment when contacted with electrode and catalyze biochemical reduction of NAD^+ at the moment when contacted with bacterial cell or enzyme. A part of NR may be contacted with electrode or bacterial cell in water-based reactant but most of that is dissolved or dispersed in the reactant. In order to induce the effective electrochemical and biochemical reaction in the bacterial culture, NR and bacterial cells have to contact continuously and simultaneously with electrode. This can be accomplished by immobilization of NR in graphite felt electrode based on the data that most of bacterial cells tend to build biofilm spontaneously on surface of solid material and the graphite felt is matrix composed of $0.47m^2/g$ of fiber (Park et al., 1999). The amino group of NR can bind covalently to alcohol group of polyvinyl alcohol by dehydration reaction, in which polyvinyl-3-imino-7-dimethylamino-2-methylphenazine (polyvinyl-NR) is produced as shown in Fig 2. The polyvinyl-NR is a water-insoluble solid electron mediator to catalyze electrochemically reduction reaction of NAD^+ like the water-soluble NR (Park and Zeikus, 2003). The polyvinyl-NR immobilized in graphite felt (NR-graphite) functions as a cathode for electron-driving circuit, an electron mediator for conversion of electric energy to electrochemical reducing power, and a catalyst for reduction of NAD^+ to NADH. The electrochemical bioreactor equipped with the NR-graphite cathode is very useful to cultivate autotrophic bacteria that grow with carbon dioxide as a sole carbon source and electrochemical reducing power as a sole energy source (Lee and Park, 2009).

3. Separation of electrochemical redox reaction

The biochemical reducing power can be regenerated electrochemically by NR-graphite cathode (working electrode) that functions as a catalyst, for which H_2O has to be electrolyzed on the surface of anode (counter electrode) that functions as an electron donor. The working electrode is required to be separated electrochemically from the counter electrode by specific septa that are the ion-selective Nafion membrane (Park and Zeikus, 2003; Kang et al., 2007; Tran et al., 2009), the ceramic membrane (Park and Zeikus, 2003; Kang et al., 2007; Tran et al., 2009), the modified ceramic membrane with cellulose acetate film (Jeon et al, 2009B), and the micro-pored glass filter, by which the electrochemical reducing power in the cathode compartment can be maintained effectively. Jeon and Park (2010) developed a combined anode that was composed of cellulose acetate film, porous ceramic membrane and porous carbon plate as shown in Fig 3. The combined anode functions as a septum for electrochemical redox separation between anode and cathode, an anode for electron-driving circuit, and a catalyst for electrolysis of H_2O. The major function of anode is to supply electrons required for generation of electrochemical reducing power in the working electrode (NR-graphite cathode), in which H_2O functions as an electron donor. The strict anaerobic bacteria that are methanogens, sulfidogens, and anaerobic fermenters grow in the condition with lower oxidation-reduction potential than -300 mV (vs. NHE) (Ferry, 1993; Gottschalk, 1985), which can be induced electrochemically inside of the carbon fibre matrices of NR-cathode under only non-oxygen atmosphere. The NR-cathode can catalyze biochemical regeneration of NADH and generation of hydrogen but can't catalyze scavenging of oxygen and oxygen radicals at around 25°C and 1 atm. The combined anode can protect effectively contamination of catholyte with the atmospheric oxygen by unidirectional evaporation of water from catholyte to atmosphere through the combined anode as shown in Fig 4. The driving force for the unidirectional evaporation of water may be generated naturally by the difference of water pressure between catholyte and outside atmosphere (Jeon et al., 2009A).

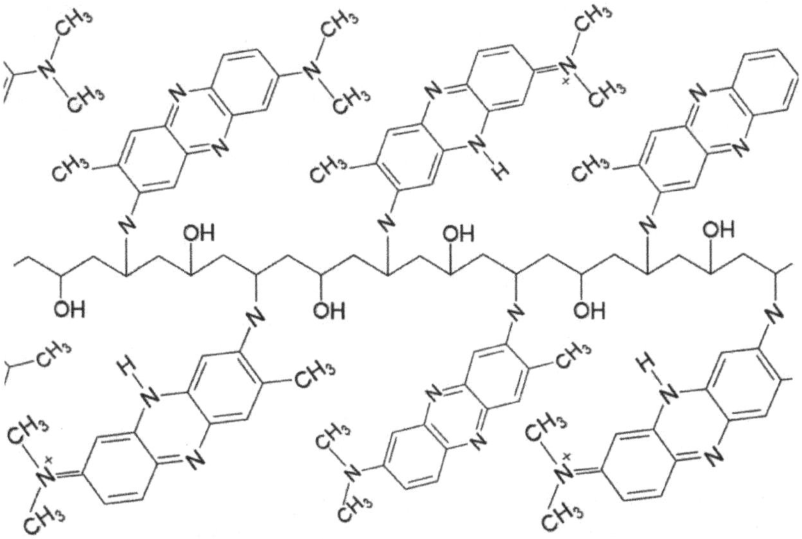

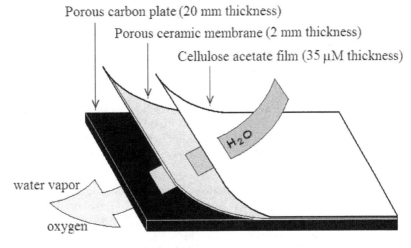

Fig. 2. Schematic structure of polyvinyl-NR that is produced by covalent bond between amine of NR and alcohol of polyvinyl alcohol. The polyvinyl-NR can bind physically to graphite cathode surface.

Fig. 3. Schematic structure of a combined anode composed of cellulose acetate film, porous ceramic membrane and porous carbon plate. Water or gas can penetrate across the cellulose acetate film but solutes can't.

Practically, the hydrogenotrophic methanogens are useful microorganisms for carbon dioxide fixation using the electrochemical bioreactor. However, most of the reducing power that is electrochemically generated in the NR-graphite cathode may be consumed to maintain the proper oxidation-reduction potential for growth of the hydrogenotrophic methanogens in the condition without chemical reducing agent. This may be a cause to decrease the regeneration effect of the biochemical reducing power and free energy in the electrochemical bioreactors. In natural ecosystem, hydrogen sulfide produced metabolically by sulfidogens in coupling with oxidation of organic acids functions as the chemical reducing agent to maintain the proper environmental condition for growth of the methanogens (Thauer et al., 1977; Oremland et al., 1989; Zinder et al., 1984).

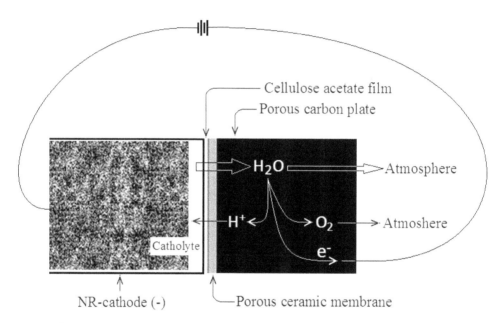

Fig. 4. Schematic structure of the combined anode composed of cellulose acetate film, porous ceramic membrane, and porous carbon plate, in which protons, electrons, and oxygen generated from water by the electrolysis may be transferred separately to the catholyte, the NR-cathode, and the atmosphere. Water is transferred from catholyte to atmosphere through the combined anode by difference of water pressure between catholyte and atmosphere.

Meanwhile, the growth condition for facultative anaerobic mixotrophs is not required to be controlled electrochemically because the metabolic function of the facultative anaerobic mixotrophs is not influenced critically by the oxidation-reduction potential. Accordingly, the combined anode may be replaced by the glass filter (pore, 1-1.6 μm) that permits transfer of water and diffusion of ions and soluble compounds. Water transferred from catholyte to anolyte through the glass filter by difference of pressure and volume is electrolysed into oxygen, protons, and electrons in the anode compartment. The protons, electrons, and oxygen are transferred separately to the catholyte, the NR-cathode, and the atmosphere as shown in Fig 5. The water in the anode compartment equipped at the center of catholyte is consumed continuously by electrolysis and refilled spontaneously from catholyte by difference of volume and pressure between the catholyte and anolyte.

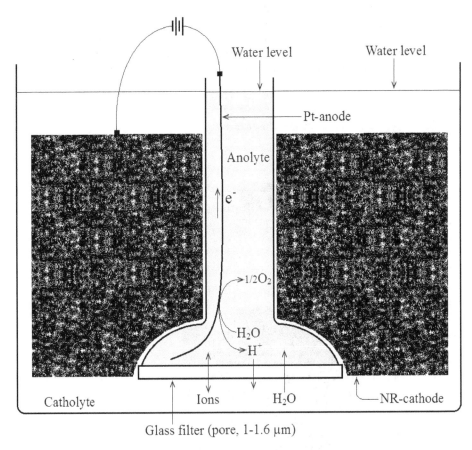

Fig. 5. Schematic structure of the anode and cathode compartment separated by glass filter. Protons, electrons, and oxygen generated from water by the electrolysis may be transferred separately to the catholyte, the NR-cathode, and the atmosphere.

4. Enrichment of hydrogenotrophic methanogens

A specially designed electrochemical bioreactor is composed of the combined anode (Fig 4) and NR-graphite cathode for enrichment of the hydrogenotrophic methanogens as shown in Fig 6. Oxygen-free and carbonate-saturated wastewater was supplied continuously from a wastewater reservoir as shown in Fig 7. The electrochemical bioreactor was operated with the electricity generated from the solar panel. The wastewater obtained from sewage treatment plant was used without sterilization, to which 50 mM of sodium bicarbonate was added. The contaminated oxygen was consumed by bacteria growing intrinsically in the wastewater reservoir. Hydrogenotrophic methanogens grow with the free energy and reducing power generated by the coupling redox reaction of carbon dioxide and hydrogen (Ferguson and Mah, 1983; Na et al., 2007; Zeikus and Wolfe, 1972). Hydrogen generated from the electrolysis of water can't function to maintain the proper oxidation reduction potential for methanogenic bacteria in the electrochemical bioreactor owing to the micro-solubility. The micro-pore formed by the fiber matrices of NR-graphite cathode may be proper micro-environment for the growth of hydrogenotrophic methanogens because hydrogen generated from NR-graphite cathode may be captured in the micro-pores and the lower oxidation-reduction potential than -300 mV (vs. NHE) may be maintained by the electrochemical reducing power.

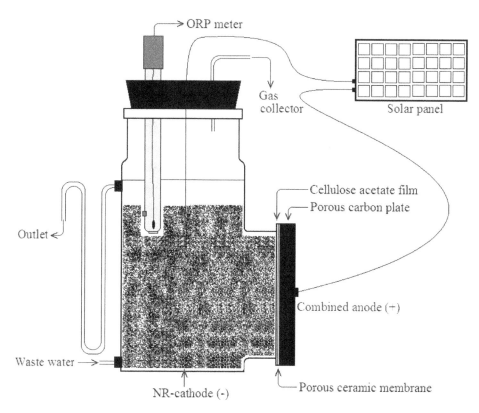

Fig. 6. Schematic structure of an electrochemical bioreactor, in which the anode compartment was replaced with the combined anode composed of cellulose acetate film, porous ceramic membrane and porous carbon plate. Water is electrolyzed in the porous carbon plate and separated into proton, electron, and oxygen.

Methyl compounds, hydrogen, low molecular weight fatty acids, hydrogen, and carbon dioxide are produced by various fermentation bacteria in the anaerobic digestive sludge. The methanogens grow syntrophically in the bioreactor cultivating anaerobic digestive sludge, which is composed of various organic compounds and anaerobic bacterial community (Stams et al., 2009; Katsuyam et al., 2009). When the anaerobic digestive sludge was applied to the electrochemical bioreactor (Fig 6 and 7), the hydrogenotrophic methanogens that are *Methanobacterium* sp., *Methanolinea* sp., and *Methnoculleus* sp. were enriched predominantly (Jeon et al., 2009B). The predominated hydrogenotrophic methanogens consumed and produce actively carbon dioxide and methane, respectively, using the electrochemical reducing power generated from the solar panel (Cheng et al., 2011). Practically, the methane production and carbon dioxide consumption were significantly increased in the electrochemical bioreactor as shown in Fig 8.

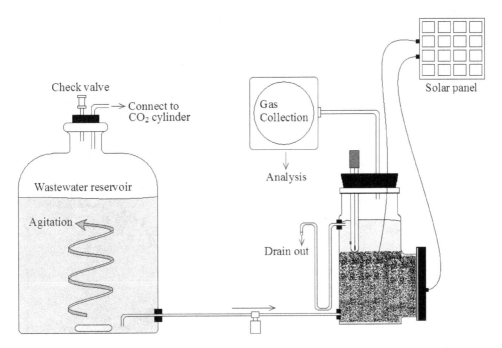

Fig. 7. Schematic structure of an electrochemical bioreactor for continuous culture of hydrogenotrophic methanogens. The wastewater saturated with carbon dioxide is supplied continuously to the electrochemical bioreactor and headspace of wastewater reservoir was refilled continuously with pure carbon dioxide without oxygen contamination.

Bacteriological conversion of carbon dioxide to methane using the electrochemical reducing power may be a technique for fixation of carbon dioxide without combustion of fossil fuel; however, may not be a way for long term storage of carbon. Cell structures of bacteria are composed of peptidoglycan, phospholipid, proteins, nucleic acids, and carbohydrates that are biochemically stable polymers (Caldwell, 1995). Bacterial cells themselves can be the carbon storage by freezing or drying without the specific engineering process. Hydrogenotrophic methanogens may not be proper carbon storage because they consume the reducing power and free energy ineffectively to maintain the lower oxidation-reduction potential than -300 mV, grow more slowly than other autotrophic bacteria, and produce the unstable metabolite (methane).

Facultative anaerobic mixotrophs, on the other hand, not only grow heterotrophically in the condition with organic carbons but also grow autotrophically in the condition with electron donors and carbon dioxide (Johson, 1998; Morikawa and Imanaka, 1993). The metabolism and physiological function of the facultative anaerobic mixotrophs are not influenced by oxygen. These are useful character of the facultative anaerobic mixotrophs to cultivate with wastewater containing other reduced organic and inorganic compounds and exhaust containing carbon dioxide, and the electrochemical reducing power as the electron donor (Skirnisdottir et al., 2001).

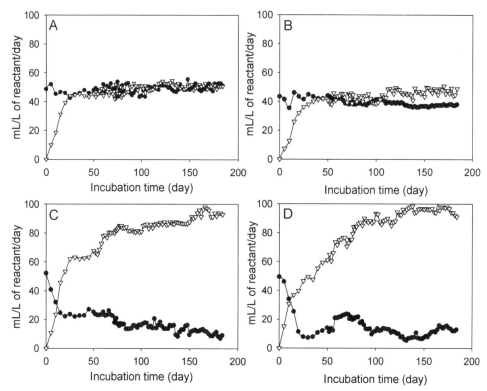

Fig. 8. Carbon dioxide consumption (●) and methane production (▽) by anaerobic digestive sludge cultivated in conventional bioreactor (reactors A and B) and electrochemical bioreactor (reactors C and D). Duplicate reactors were operated to enhance the comparability between the conventional bioreactor and the electrochemical bioreactor.

5. Enrichment and cultivation of carbon dioxide-fixing bacteria

A cylinder-type electrochemical bioreactor composed of the built-in anode compartment and NR-graphite cathode was employed to enrich the facultative anaerobic mixotrophs capable of fixing carbon dioxide with electrochemical reducing power as shown in Fig 9. The NR-cathode was separated electrochemically from anode compartment by the glass filter (Fig 5). Mixture of the bacterial community obtained from aerobic wastewater treatment reactor, forest soil, and anaerobic wastewater was cultivated in the cylinder-type electrochemical bioreactor to enrich selectively carbon dioxide-fixing bacteria with the electrochemical reducing power generated from NR-graphite cathode. DC -3 volt of electricity that was generated by a solar panel was charged to NR-graphite cathode to induce generation of electrochemical reducing power. Electricity is the easiest energy to transfer and supply to any electronic device. The electrochemical bioreactor is also the simplest electronic device to convert electric energy to biochemical reducing power. The wastewater and exhausted gas can be used directly without purification or separation as the nutrient source for bacterial metabolism. Experimentally, the electricity generated from the 25 cm² of the solar panel is very enough for operation of the 10 L of electrochemical bioreactor.

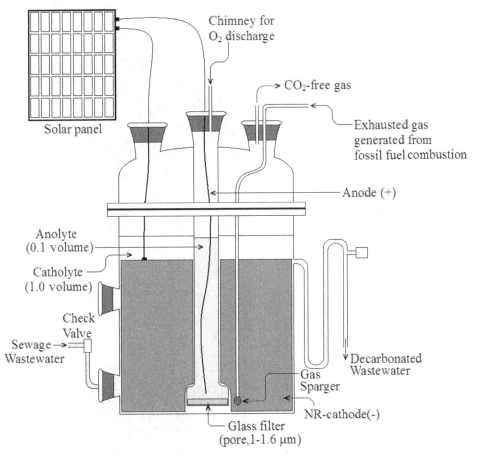

Fig. 9. Schematic diagram of the cylinder-type electrochemical bioreactor equipped with a built-in anode compartment for the cultivation of CO_2-fixing bacteria. The glass filter septum equipped at the bottom end of the anode compartment functions as redox separator between anode and cathode compartment and micropore for transfer of catholyte to anode compartment.

During enrichment of the carbon dioxide-fixing bacteria using the cylinder-type electrochemical bioreactor, bacterial community was changed significantly as show in Fig 10. Some of bacteria community was increased or enriched as shown in the box A and C but decreased or died out as shown in the box B and D. These phenomena are a clue that the bacterial species that can fix carbon dioxide with electrochemical reducing power are adapted selectively to the reactor condition but other bacteria that can't generate biochemical reducing power from the electrochemical reducing power are not. The DNA bands were extracted from TGGE gel and sequenced. Identity of the bacteria was determined based on the 16S-rDNA sequence homology.

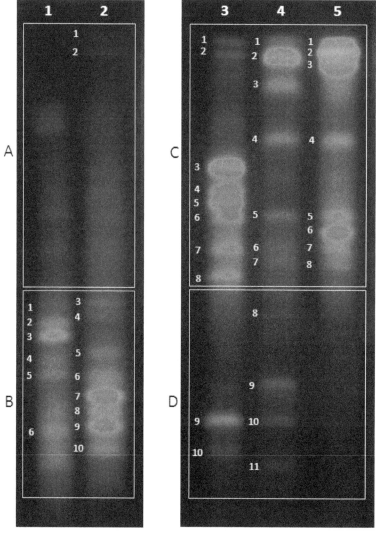

Fig. 10. TGGE patterns of 16S-rDNA variable regions amplified with chromosomal DNA extracted from bacterial communities enriched in the cylinder-type electrochemical bioreactors. 50 ml of bacterial culture was isolated from the electrochemical bioreactor at the initial time immediately after inoculation (lane1), 2nd week (lane 2), 8th week (lane 3), 16th week (lane 4), and 24th week of incubation time (lane 5).

Lane	Band	Genus or Species	Homology (%)	Accession No.
1 (initial)	1	Uncultured *Burkholderia* sp.	98	FJ393136
	2	Groundwater biofilm bacterium	98	FJ204452
	3	*Hydrogenophaga* sp.	98	FM998722
	4	Uncultured bacterium sp.	97	HM481230
	5	*Aquamicrobium* sp.	98	GQ254286
	6	Uncultured *Actinobacterium* sp.	99	FM253013
2 (2nd week)	1	Uncultured bacterium sp.	97	AF234127
	2	Uncultured bacterium sp.	97	EU532796
	3	Uncultured *Clostridum* sp.	99	FJ930072
	4	Uncultured *Polaromonas* sp.	99	HM486175
	5	Uncultured *Rhizobium* sp.	100	FM877981
	6	*Raoultella planticola*	98	EF551363
	7	Unidentified bacterium	98	AV669107
	8	Uncultured bacterium	99	HM920740
	9	Uncultured bacterium	97	GQ158957
	10	Uncultured *Klebsiella* sp.	98	GQ416299
3 (8th week)	1	Uncultured bacterium sp.	97	AF234127
	2	Uncultured bacterium sp.	97	EU532796
	3	*Enterococcus* sp.	98	DQ305313
	4	Uncultured bacterium	98	HM820223
	5	*Aerosphaera taera*	99	EF111256
	6	*Alcaligenes* sp.	98	GQ383898
	7	Uncultured bacterium	98	HM231340
	8	Uncultured bacterium sp.	97	FJ675330
	9	*Stenotrophomonas* sp.	98	EU635492
	10	Uncultured *Klebsiella* sp.	98	GQ416299
4 (16th week)	1	Uncultured bacterium sp.	97	AF234127
	2	Uncultured bacterium sp.	97	EU532796
	3	Uncultured bacterium sp.	98	HM575088
	4	*Alcaligenes* sp.	98	GQ200556
	5	*Alcaligenes* sp.	98	GQ383898
	6	Uncultured bacterium	98	HM231340
	7	*Achromobacter* sp.	96	GQ214399
	8	Uncultured *Lactobacillales* bacterium sp.	96	HM231341
	9	Uncultured *Ochrombacterum* sp.	97	EU882419
	10	*Stenotrophomonas* sp.	98	EU635492
	11	*Tissierella* sp.	96	GQ461822
5 (24th week)	1	Uncultured bacterium sp.	97	AF234127
	2	Uncultured bacterium sp.	97	EU532796
	3	Uncultured bacterium sp.	98	HM820116
	4	*Alcaligenes* sp.	98	GQ200556
	5	*Alcaligenes* sp.	97	GQ383898
	6	*Enterococcus sp.*	99	FJ513901
	7	Uncultured bacterium	98	HM231340
	8	*Achromobacter* sp.	96	GQ214399

Table 1. The homologous bacterial species with the sequences of DNA extracted from TGGE bands (Fig 10), which were identified based on the GenBank database.

Some anaerobic bacteria (*Hydrogenophaga* sp. and *Clostridium* sp.) that may be originated from the anaerobic wastewater treatment reactor are detected at the initial cultivation time but disappeared after 8th week of incubation time (Kang and Kim, 1999; Willems et al., 1989; Lamed et al., 1988). On the other hand, the bacteria that are capable of fixing carbon dioxide by autotrophic or mixotrophic metabolism were enriched as shown in Table 1. All of the enriched bacteria may not be the carbon dioxide-fixing bacteria but *Achromobacter* sp. and *Alcaligenes* sp. are known to fix carbon dioxide autotrophically or mixotrophically (Freter and Bowien, 1994: Friedrich, 1982; Hamilton et al., 1965; Leadbeater and Bowien, 1984; Ohmura et al.). During the enrichment of the carbon dioxide-fixing bacteria, carbon dioxide consumption was increased and reached to stationary phase after 15th week of incubation time as shown in Fig 11. Various organic compounds contained in the bacterial cultures that were originated from anaerobic wastewater treatment reactor, aerobic wastewater treatment reactor, and forest soil might be consumed completely and then carbon dioxide-fixing bacteria might grow selectively. The carbon dioxide consumption was increased initially and then reached to stationary phase after 15th week of incubation time, which is proportional to the enrichment time of the *Achromobacter* sp. and *Alcaligenes* sp.

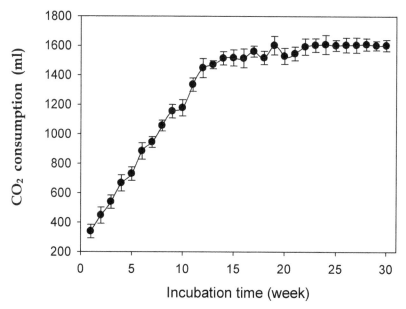

Fig. 11. Weekly consumption of CO_2 in the electrochemical bioreactor from the initial incubation time to 30 weeks. CO_2 consumption was analyzed weekly and the gas reservoir was refilled with 50±1% of CO_2 to N_2 at 4-week intervals.

Before and after enrichment, the bacterial community grown in the cylinder-type electrochemical bioreactor was analyzed using the pyrosequencing technique (Van der Bogert et al., 2011). The classifiable sequences obtained by the pyrosequencing were identified based on the Ribosomal Database Project (RDP), and defined at the 100 % sequence homologous level. The most abundant sequences (17.96%) obtained from the bacterial culture before enrichment was identified as *Brevundimonas* sp., and the abundance of sequences identified with *Alcaligenes* sp. and *Achromobcter* sp. was 0.98 and 0.12%, respectively. Meanwhile, the most abundant sequences (43.83%) obtained from the bacterial culture after enrichment was identified as *Achromobacter* sp., and the most classifiable sequences were also identified as *Achromobacter* sp. and *Alcaligenes* sp. as shown in Table 2.

Before enrichment				After enrichment			
Classifiable sequences	Abundance (%)	Bacterial genus	Homology (%)	Classifiable sequences	Abundance (%)	Bacterial genus	Homology (%)
876	17.96	*Brevundimonas*	100	2248	43.83	*Achromobacter*	100
153	3.14	*Pseudomonas*	100	748	14.58	*Achromobacter*	100
111	2.28	*Hydrogenophaga*	100	595	5.87	*Stenotrophomonas*	100
99	2.03	*Delftia*	100	301	2.28	*Achromobacter*	100
86	1.76	*Stenotrophomonas*	100	263	1.77	*Achromobacter*	100
70	1.44	*Pseudomonas*	100	219	1.23	*Achromobacter*	100
53	1.09	*Parvibaculum*	100	117	0.90	*Achromobacter*	100
52	1.07	*Brevundimonas*	100	91	0.66	*Achromobacter*	100
48	0.98	*Alcaligenes*	100	63	0.57	*Alcaligenes*	100
32	0.66	*Comamonas*	100	46	0.53	*Achromobacter*	100
31	0.64	*Bacillus*	100	34	0.49	*Achromobacter*	100
26	0.53	*Bosea*	100	29	0.49	*Castellaniella*	100
21	0.43	*Devosia*	100	27	0.45	*Achromobacter*	100
17	0.35	*Acidovorax*	100	25	0.45	*Achromobacter*	100
12	0.25	*Brevundimonas*	100	25	0.39	*Stenotrophomonas*	100
12	0.25	*Sphaerobacter*	100	23	0.16	*Achromobacter*	100
11	0.23	*Brevundimonas*	100	23	0.14	*Alcaligenes*	100
9	0.18	*Acinetobacter*	100	20	0.12	*Achromobacter*	100
9	0.18	*Sphaerobacter*	100	14	0.10	*Alcaligenes*	100
8	0.16	*Brevundimonas*	100	14	0.10	*Pseudomonas*	100
7	0.14	*Hyphomicrobium*	100	11	0.08	*Achromobacter*	100
7	0.14	*Thermomonas*	100	10	0.08	*Achromobacter*	100
6	0.12	*Achromobacter*	100	8	0.06	*Achromobacter*	100
6	0.12	*Brevundimonas*	100	7	0.06	*Achromobacter*	100
4	0.10	*Devosia*	100	7	0.06	*Achromobacter*	100
3	0.08	*Pseudoxanthomonas*	100	6	0.04	*Alcaligenes*	100
3	0.06	*Castellaniella*	100	6	0.04	*Achromobacter*	100
3	0.06	*Gordonia*	100	6	0.04	*Achromobacter*	100

Table 2. Relative abundances of dominant bacterial taxa in the bacterial culture before and after enrichment. The relative abundances were estimated from the proportion of classifiable sequences, excluding those sequences that could not be classified below the genus level and 100% homology with the specific bacterial genus.

The *Achromobacter* sp. described in previous research was a facultative chemoautotroph (Hamilton *et al.*, 1965; Romanov *et al.*, 1977); however, it grew autotrophically with electrochemical reducing power under a CO_2 atmosphere and consumed CO_2 in this study. This result demonstrates that *Achromobacter* sp. grown in the electrochemical bioreactor may be a chemoautotroph capable of fixing CO_2 with the electrochemical reducing power. Meanwhile, various articles have reported that *Alcaligenes* sp. grew autotrophically (Frete and Bowien, 1994; Doyle and Arp. 1987; Leadbeater and Bowien, 1984) or heterotrophically (Reutz *et al.*, 1982). According to these articles, *Alcaligenes* spp. are capable of growing autotrophically with a gas mixture of H_2, CO_2, and O_2, as well as heterotrophically under air on a broad variety of organic substrates. *Alcaligenes* spp. metabolically oxidize H_2 to regenerate the reducing power during autotrophic growth under H_2-CO_2 atmosphere (Hogrefe *et al.*, 1984). The essential requirement for the autotrophic growth of both *Achromobacter* spp. and *Alcaligenes* spp. under CO_2 atmosphere is to regenerate reducing power in conjunction with metabolic H_2 oxidation, which may be replaced by the electrochemical reducing power on the basis of the results obtained in this research. The electrochemical reducing power required for the cultivation of carbon-dioxide fixing bacteria can be produced completely by the solar panel, by which atmospheric carbon dioxide may be fixed by same system to the photosynthesis.

6. Strategy of atmospheric carbon dioxide fixation using the solar energy

In global ecosystem, land plants, aquatic plants, and photoautotrophic microorganisms produce biomass that is original source of organic compounds (O'Leary, 1988). Autotrophs that are growing naturally or cultivating artificially have fixed the atmospheric carbon dioxide generated by heterotrophs, by which the atmospheric carbon dioxide may be balanced ecologically. However, the carbon dioxide generated from the combustion of organic compounds (petroleum and coal) that are not originated from biomass may be accumulated additionally in the atmosphere, inland water, and sea water. The solar radiation that reaches to the earth may not be limited for photosynthesis of phototrophs or electric generation of solar cells; however, the general habitats for growth of the phototrophs have been decreased by various human activities and the places for installation of the solar cells are limited to the habitats for human. If the solar cells were installed in the natural habitats, phototrophic fixation of carbon dioxide may be decreased in proportion to the electricity generation by the solar cells. The constructions of new cities, farmlands, golf courses, ski resorts, and sport grounds cause to convert the forests to grass field whose ability for carbon dioxide fixation is greatly lower than the forest. Consequently, the plantation of trees and grasses in the habitable lands or cultivation of algae and cyanobacteria in the habitable waters can't be the way to decrease additionally the atmospheric carbon dioxide.

Carbon dioxide has been fixed biologically by photoautotrophic, chemoautotrophic and mixotrophic organisms. The photoautotrophic bacteria assimilate carbon dioxide into organic compounds for cell structures with reducing power regenerated by the solar radiation under atmospheric condition (Kresge et al., 2005). The chemoautotrophs assimilate carbon dioxide into cell structure in coupling with production of methane or acetic acid with reducing power regenerated by hydrogenase under strict anaerobic hydrogen atmosphere (Perreault et al., 2007). The mixotrophs assimilate carbon dioxide into biomolecules with reducing power regenerated in coupling with metabolic oxidation of organic or inorganic compounds (Eiler, 2006). The photoautotrophs, chemoautotrophs, and mixotrophs can reduce metabolically carbon dioxide to organic carbon with the common reducing power (NADH or NADPH), which, however, are regenerated by different metabolisms. The photoautotrophs, especially cyanobacteria that fix carbon dioxide by completely same metabolism (Calvin cycle) with plants, appear as if they are ideal organism to fix biologically carbon dioxide without chemical energy; however, they are unfavorable to be cultivated in the tank-type bioreactor owing to the limitation of reachable distance of solar radiation in aquatic condition. The chemoautotrophs may be useful to produce methane and acetic acid from carbon dioxide; however, they can grow only in the limit condition of the lower redox potential than -300 mV (vs. NHE) and with hydrogen. The mixotrophs can grow in the condition with electron donors, which are regardless of organic or inorganic compounds, for regeneration of reducing power under aerobic and anaerobic condition. This is the reason why the facultative anaerobic mixotrophs may be more effective than others to fix the atmospheric carbon dioxide directly by simple process. Especially, the cylinder-type electrochemical bioreactor equipped with the built-in anode compartment (Fig 9) is an optimal system for the cultivation or enrichment of facultative anaerobic mixotrophs. Basements of buildings or villages are used generally for maintenances or facilities for wastewater collection, electricity distribution, tap water distribution, and garage. The basements can't be the habitats for cultivation of plants with the natural sun light but can be utilized for cultivation of the carbon dioxide-fixing bacteria with electric energy generated from the solar cells that can be installed on the rooftop as shown in Fig 12.

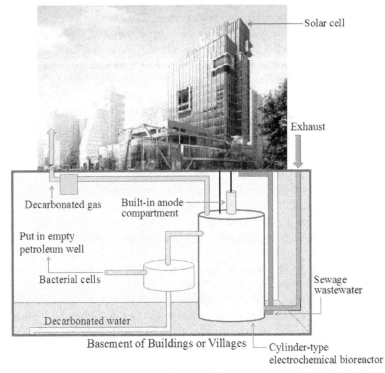

Fig. 12. Schematic structure of the electrochemical bioreactors installed in the building basement. The carbon dioxide-fixing bacteria can be cultivated using the electric energy generated by the solar cells.

The facultative anaerobic mixotrophs assimilate heterotrophically organic compounds contained in the wastewater into the structural compounds of bacterial cells under oxidation condition but autotrophically carbon dioxide into the biomass under condition with high balance of biochemical reducing power (NADH/NAD$^+$). DC electricity generated from the solar cells can be transferred very conveniently to the cylinder-type electrochemical bioreactor without conversion, which is the energy source for increase of biochemical reducing power balance. A part of the atmospheric carbon dioxide has been generated from the combustion system of fossil fuel, which may be required to be return to the empty petroleum well. To store the bacterial cells in the empty petroleum well is to return the carbon dioxide generated from petroleum combustion to the original place. The peptidoglycans, phospholipids, proteins, and nucleic acids that are major ingredients of bacterial cell structures are stable chemically to be stored in the empty petroleum well owing to the non-oxygenic condition. Conclusively, what the atmospheric carbon dioxide originated from the petroleum and coal is returned to the original place again may be best way to decrease the greenhouse effect.

7. Conclusion

The atmospheric carbon dioxide originated from petroleum and coal is required to be completely isolated from the ecological material cycles. The carbons in the ecological system are accumulated as the organic compounds in the organisms and as the carbon dioxide in the atmosphere, which is cycled via the photosynthesis and respiration, especially, plants are the biggest pool for carbon storage. However, the forest and plant-habitable area has been decreased continuously by human activities.

The cultivation of cyanobacteria and single cell algae with solar energy may be the best way to isolated effectively carbon dioxide from atmosphere but is possible in the water pool-type reactor located in the plant-habitable area. In other words, the forests or grass lands may be replaced by the water pools, by which the effect of carbon dioxide fixation has to be decreased. The cyanobacteria and algae can be cultivated in the bioreactor using lamp light operated with electric energy that is generated from solar cells, for which the solar energy has to be converted to electric energy and then converted again to the light energy. These phototrophic microorganisms have been studied actively and applied to produce nutrient sources and pharmacy. The goal for cultivation of the phototrophic microorganisms is to produce the utilizable materials but not to fix carbon dioxide like the agricultural purpose.

The carbon compounds of the organic nutritional compounds contained in the sewage wastewater are the potential carbon dioxide, which may be the useful medium for cultivation of the mixotrophic bacteria capable of fixing carbon dioxide. The maximal balance of anabolism to catabolism is theoretically 0.4 to 0.6 in the mixotrophic bacteria growing with organic carbons as the energy source, in which the carbon dioxide can't be the source for both anabolism and catabolism; however, the balance can be changed by the external energy like the electrochemical reducing power. In the condition with both the organic carbons and the electrochemical reducing power as the energy source, the balance of anabolism to catabolism may be increased to be higher than 0.4 due to the carbon dioxide assimilation that is generated in coupling with the redox reaction of biochemical reducing power electrochemically regenerated. The electrochemical reducing power can induce regeneration of NADH and ATP, by which both the assimilation of organic carbon and carbon dioxide into bacterial structure compounds can be activated. The goal of cultivation of bacterial cells using the cylinder-type electrochemical is to assimilate the atmospheric carbon dioxide to the organic compounds for bacterial structure without the combustion of fossil fuel and without production of metabolites. Some metabolites that are methane and acetic acid can be generated by the strict anaerobic bacteria under anaerobic hydrogen-carbon dioxide atmosphere but not useful for industrial utility owing to the cost for production. Meanwhile, the methane and acetic acid produced from the organic compounds in the process for treatment of wastewater or waste materials may be useful as the by-product for the industrial utility. The cell size and structural character of bacteria permits to put directly the bacterial cells in the empty petroleum well without any process, by which the atmospheric carbon dioxides are returned to the original place.

8. Acknowledgement

Writing of this chapter was supported by the New & Renewable Energy of the Korea Institute of Energy Technology Evaluation and Planning (KETEP) grant funded by the Korea government Ministry of Knowledge Economy (2010T1001100334)

9. References

Caldwell, D.R. 1995. Microbial physiology and metabolism. Pp. 5-23. Wm. C. Brown Publishers. Oxford. England.

Canadell, J.G., C.R. Quéré, M.R. Raupach, C.B. Field, E.T. Buitenhuis, P. Cialis, T.J. Conway, N.P. Gillett, R.A. Houghton, and G. Marland. 2007. Contributions to accelerating atmospheric CO_2 growth from economic activity, carbon intensity, and efficiency of natural sinks. PNAS 104: 18866-18870.

Cheng, K.Y., G. Ho, and R. Cord-Ruwisch. 2011. Novel methanogenic rotatable bioelectrochemical system operated with polarity inversion. Environ. Sci. Technol. 45: 796-802.

Cox, P.M., R.A. Betts, C.D. Jones, S.A. Spall, and I.J. Totterdell. 2000. Acceleration of global warming due to carbon-cycle feedbacks in a coupled climate model. Nature 4008: 184-187.

Cramer, W., A. Bondeau, S. Schaphoff, W. Lucht, B. Smith, and S. Sitch. 2004. Tropical forests and the global carbon cycle: impacts of atmospheric carbon dioxide, climate change and rate of deforestation. Phil. Trans. R. Soc. Lond. B 359: 331-343.

Dhillon, A., M. Lever, K.G. Lloyd, D.B. Albert, M.L. Sogin, and A. Teske. 2005. Methanogen diversity evidenced by molecular characterization of methyl coenzyme M reductase A (mcrA) genes in hydrothermal sediments of the Suaymas Basin. Appl. Environ. Microbiol. 71: 4592-4601.

Eiler, A. 2006. Evidence for the ubiquity of mixotrophic bacteria in the upper ocean: implications and consequences. Appl. Environ. Microbiol. 72: 7431-7437.

Ferguson, T.J., and R.A. Mah. 1983. Isolation and characterization of an H_2-oxidizing thermophilic methanogen. Appl. Environ. Microbiol. 45: 265-274.

Ferry, J.G. 1993. Methanogenesis: Ecology, Physiology, Biochemistry and Genetics. Chapman & Hall, New York.

Freter, A., and B. Bowien. 1994. Identification of a novel gene, aut. involved in autotrophic growth of Alcaligenes eutrophus. J. Bacteriol. 176: 5401-5408.

Friedrich, C., 1982. Derepression of hydrogenase during limitation of electron donor and derepression of ribulosebiophosphate carboxylase during carbon limitation of Alcaligenes eutrophus. J. Bacteriol. 149:203-210.

Gerlach, T.M., K.A. McGee, T. Elias, A.J. Sutton, and M.P. Doukas. 2002. Carbon dioxide emission rate of Kilauea volcano: Implications for primary magma and the summit reservoir. J. Geophys. Res. 107:2189-2203.

Gottschalk, G. 1985. Bacterial metabolism, Second Edition, Pp. 252-260. Springer-Verlag, New York.

Grulke, N.E., G.H. Riechers, W.C. Oechel, U. Hjelm, and C. Jaeger. 1990. Carbon balance in tussock tundra under ambient and elevated atmosphere. Oecologia 83: 485-494.

Hamilton, R.R., R.H. Burris, P.W. Wilson, and C.H. Wang. 1965. Pyruvate metabolism, carbon dioxide assimilation, and nitrogen fixation by an Achromobacter species. J. Bacteriol. 89:647-653.

Hansen, K. 2008. Water vapor confirmed as major in climate change. News topics from NASA. http://www.nasa.gov/topics/earth/features/vapor_warming.html

Held, I.M., and B.J. Soden. 2000. Water vapor feedback and global warming. Annu. Rev. Energ. Environ. 25: 441-475.

Hogrefe, C., D. Römermann, and B. Friedrich. 1984. Alcaligenes eutrophus hydrogenase gene (Hox). J. Bacteriol. 158, 43-48.

Jeon, B.Y., and D.H. Park. 2010. Improvement of ethanol production by electrochemical redox combination of Zymomonas mobilis and Saccharomyces cerevisiae. J. Microbiol. Biotechnol. 20: 94-100.

Jeon, B.Y., S.Y. Kim, Y.K. Park, and D.H. Park. 2009A. Enrichment of hydrogenotrophic methanogens in coupling with methane production using electrochemical bioreactor. J. Microbiol. Biotechnol. 19: 1665-1671.

Jeon, B.Y., T.S. Hwang, and D.H. Park. 2009B. Electrochemical and biochemical analysis of ethanol fermentation of Zymomonas mobilis KCCM11336. J. Microbiol. Biotechnol. 19: 666-674.

Johnson, D.B. 1998. Biodiversity and ecology of acidophilic microorganisms. FEMS Microbiology Ecology 27: 307-317.

Kang, B., and Y.M. Kim. 1999. Cloning and molecular characterization of the genes for carbon monoxide dehydrogenase and localization of molybdopterin, flavin adenine dinucleotide, and iron-sulfur centers in the enzyme of *Hydrogenophaga pseudoflava*. J. Bacteriol. 181: 5581-5590.

Kang, H.S., B.K. Na, and D.H. Park. 2007. Oxidation of butane to butanol coupled to electrochemical redox reaction of $NAD^+/NADH$. Biotech. Lett. 29: 1277-1280.

Katsuyama, C., S. Nakaoka, Y. Takeuchi, K. Tago, M. Hayatsu, and K. Kato. 2009. Complementary cooperation between two syntrophic bacteria in pesticide degradation. J. Theoretical Biol. 256: 644-654.

Keeling, C.D., T.P. Whorf, M. Wahlen, and J. Vanderplicht. 1995. Interannual extremes in the rate of rise of atmospheric carbon-dioxide since 1980, Nature. 375: 666-670.

Kiehl, J., T. Kevin, and E. Trenberth. 1997. Earth's annual global mean energy budget. Bulletin of the American Meteological Society 78: 197-208.

Lamed, R.J., J.H. Lobos, and T.M. Su. 1988. Effects of stirring and hydrogen on fermentation products of *Clostridium thermocellum*. Appl. Environ. Microbiol. 54: 1216-1221.

Lashof, D.A., and D.R. Ahuja. 1990. Relative contributions of greenhouse gas emission to global warming. Nature 344: 529-531.

Leadbeater, L., and B. Bowien. 1984. Control autotrophic carbon assimilation *Alcaligenes eutrophus* by inactivation and reaction of phosphoribulokinase. J. Bacteriol. 57: 95-99.

Lee, W.J., and D.H. Park. 2009. Electrochemical activation of nitrate reduction to nitrogen by *Ochrobactrum* sp. G3-1 using a noncompartmetned electrochemical bioreactor. J. Microbiol. Biotechnol. 19: 836-844.

McKane, R.B., E.B. Rastetter, J.M. Melillo, G.R. Shaver, C.S. Hopkinson, D.N. Femandes, D.L. Skole, and W.H. Chomentowski. 1995. Effects of global change on carbon storage in tropical forests of south America. Global Biochemical Cycle 9: 329-350.

Morikawa, M., and T. Imakawa. 1993. Isolation of a new mixotrophic bacterium which can fix aliphatic and aromatic hydrocarbons anaerobically. J. Ferment. Bioengin. 4: 280-283.

Na, B.K., T.K. Hwang, S.H. Lee, D.H. Ju, B.I. Sang, and D.H. Park. 2007. Development of bioreactor for enrichment of chemolithotrophic methanogen and methane production. Kor. J. Microbiol. Biotechnol.. 35: 52-57.

O'Leary, M.H. 1988. Carbon isotopes in photosynthesis. BioScienece 38: 328-336.

Kresge, N., R.D. Simoni, and R.L. Hill. 2005. The discovery of heterotrophic carbon dioxide fixation by Harland G Wood. J. Biol. Chem. 139:365-376.

Ohmura, N., K. Sasaki, N. Matsumoto, and H. Saiki. 2002. Anaerobic respiration using Fe^{3+}, S^o, and H_2 in the chemoautotrophic bacterium *Acidithiobacillus ferroxidans*. J. Bacteriol. 184: 2081-2087.

O'Regan, B., and M. Grätzel. 1991. A low-cost, high-efficiency solar cell based on dye-sensitized colloidal TiO_2 films. Nature 353: 737-740.

Oremland, R.S., R.P. Kiene, I. Mathrani, M.J. Whitica, and D.R. Boone. 1989. Description of an estuarine methylotrophic methanogen which grows on dimethyl sulfide. Appl. Environ. Microbiol. 55: 994-1002.

Park, D.H., and Z.G. Zeikus. 1999. Utilization of electrically reduced neutral red by *Actinobacillus succinogenes*: physiological function of neutral red in membrane-driven fumarate reduction and energy conservation. J. Bacteriol. 181: 2403-2401.

Park, D.H., and J.G. Zeikus. 2000. Electricity generation in microbial fuel cells using neutral red as an electronophore. Appl. Environ. Microbiol. 66:1292-1297.

Park, D.H., and J.G. Zeikus. 2003. Improved fuel cell and electrode designs for producing electricity from microbial degradation. Biotechnol. Bioengin. 81: 348-355.

Park, D.H., B.H. Kim, B. Moore, H.A.O. Hill, M.K. Song, and H.W. Rhee. 1997. Electrode reaction of *Desulforvibrio desulfuricans* modified with organic conductive compounds. Biotech. Technique. 11: 145-148.

Park, D.H., M. Laiveniek, M.V. Guettler, M.K. Jain, and J.G. Zeikus. 1999. Microbial utilization of electrically reduced neutral red as the sole electron donor for growth of metabolite production. Appl. Environ. Microbiol. 2912-2917.

Perreault, N.N., C.W. Greer, D.T. Andersen, S. Tille, G. Lacrampe-Couloume, B. Sherwood, and L.G. Whyte. 2008. Heterotrophic and autotrophic microbial populations in cold perennial springs of the high arctic. Appl. Environ. Microbiol. 74: 6898-6907.

Petty, G.W. 2004. A first course in atmospheric radiation, pp. 29-251, Sundog Publishing.

Pollack, J.D., J. Banzon, K. Donelson, J.G. Tully, J.W. Davis, K.J. Hackett, C. Agbayyim, and R.J. Miles. 1996. Reduction of benzyl viologen distinguishes genera of the class *Mollicutes*. Int. J. System. Bacteriol. 46: 881-884.

Reutz, I., P. Schobert, and B. Bowien. 1982. Effect of phosphoglycerate mutase deficiency on heterotrophic and autotrophic carbon metabolism of *Alcaligenes eutrophus*. J. Bacteriol. 151: 8-14.

Raich, J.W., and W.H. Schlesinger. 2002. The global carbon dioxide flux in soil respiration and its relationship to vegetation and climate. Tellus 44:81-99.

Robertson, G.P., and J.M. Tiejei. 1988. Deforestation alters denitrification in a lowland tropical rain forest. Nature 136: 756-759.

Romanova, A.K., A.V. Nozhevnikova, J.G. Leonthev, and S.A. Alekseeva. 1977. Pathways of assimilation of carbon oxides in *Seliberia carboxydohydrogena* and *Achromobacter carboxydus*. Microbiology 46, 719-722.

Schmidt, G.A., R. Ruedy, R.L. Miller, and A.A. Lacis. 2010. Attribution of the present-day total greenhouse effect. J. Geophys. Res. 115, D20106, doi:10.1029/2010JD014287.

Shine, K.P., M. Piers, and de F. Forster. 1999. The effect of human activity on radiative forcing of climate change: a review of recent developments. Global and Planetary Change 20:205-225.

Skirnisdottir, S., G.O. Hreggvidsson, O. Holst, and J.K. Kristjansson. 2001. Isolation and characterization of a mixotrophic sulfur-oxidizing *Thermus scotoductus*. Extremophiles. 5: 45-51.

Skole, D.L., W.A. Salas, and C. Silapathong. 1998. Interannual variation in the terrestrial carbon cycle: significance of Asian tropic forest conversion to imbalanced in the global carbon budget. Pp. 162-186 in J.N Galoway and J.M. Melillo, eds. Asian change in the context of global change. Cambridge: Cambridge University Press.

Stams, A.J.M., and C.M. Plugge. 2009. Electron transfer in syntrophic communities of anaerobic bacteria and archaea. Nature Rev. Microbiol. 7: 568-577.

Stork, N.E. 1977. Measuring global biodiversity and its decline. Pp. 46-68 in M.L. Reaka-Kudia at al., eds. Biodiversity LL: Understanding and protecting our natural resources. Washington, DC: Joseph Henry Press.

Tans, P. 2011. "Trends in atmospheric carbon dioxide". National Oceanic & Atmospheric Administration, Earth system Research Laboratory of Global Monitoring Division. Retrieved 2011-01-19.

Thauer, R.K., K. Jungermann, and K. Decker. 1977. Energy conservation in chemotrophic anaerobic bacteria. Bacteriol. Rev. 41: 100-180.

Tran, H.T., D.H. Kim, S.J. Oh, K. Rasool, D.H. Park, R.H. Zhang, and D.H. Ahn. 2009. Nitrifying biocathode enable effective electricity generation and sustainable wastewater treatment with microbial fuel cell. Water Sci. Technol. 59: 1803-1808.

Van der Bogert, B., W.M. de Vos, E.G. Zoetendal, and M. Kleerevezem. 2011. Microarray analysis and barcoded pyrosequencing provide consistent microbial profiles. Appl. Environ. Microbiol. 77: 2071-2080.

Van Veen, J.A., E. Liljeroth, and J.J.A. Lekkerkerk. 1991. Carbon fluxes in plant-soil systems at elevated atmospheric CO2 levels. Ecol. Appl. 1: 175-181.

Wang, S., and D. Du. 2002. Studies on the electrochemical behavior of hydroquinone at L-cysteine self-assembled monolayers modified gold electrode. Sensors 2: 41-49.

Willems, A., J. Busse, M. Goor, B. Pot, E. Falsen, E. Jantzen, B. Hoste, M. Gillis, K. Kersters, G. Auling, and J. Delay. 1989. *Hydrogenophaga*, a new genus of hydrogen-oxidizing bacteria that includes *Hydrogenophaga flava* comb. nov. (formerly *Pseudomonas flava*). *Hydrogenophaga palleronii* (formerly *Pseudomonas palleroni*), *Hydrogenophaga pseudoflava* (formerly *Pseudomonas pseudoflava* and "*Pseudomonas carboxydoflava*"), and *Hydrogenophaga taeniospiralis* (formerly *Pseudomonas taeniospiralis*). Int. J. Syst. Bacteriol. 39: 319-333.

Williams, S.N., S.J. Schaefer, V.M. Lucia, and D. Lopez. 1992. Global carbon dioxide emission to the atmosphere by volcanoes. Geochim. Cosmochim. Acta. 56: 1765-1770.

Worrell, E., L. Price, N. Martin, C. Hendriks, and L.O. Meida. 2001. Carbon dioxide emissions from the global cement industry. Ann. Rev. Energy Environ. 26: 303-329.

Zeikus, J.G., and R.S. Wolfe. 1972. *Methanobacterium thermoautotrophicum* sp. nov.: An anaerobic autotrophic, extreme thermophile. J. Bacteriol. 109: 707-713.

Zinder, S.H., S.C. Cardwell, T. Anguish, M. Lee, and M. Koch. 1984. Methanogenesis in a thermophilic (58°C) anaerobic digester: *Methanothrix* sp. as an important aceticlastic methanogen. Appl. Environ. Microbiol. 47: 796-807.

9

Cuprous Oxide as an Active Material for Solar Cells

Sanja Bugarinović[1], Mirjana Rajčić-Vujasinović[2],
Zoran Stević[2] and Vesna Grekulović[2]
[1]IHIS, Science and Technology Park "Zemun", Belgrade,
[2]University of Belgrade, Technical faculty in Bor, Bor
Serbia

1. Introduction

Growing demand for energy sources that are cleaner and more economical led to intensive research on alternative energy sources such as rechargeable lithium batteries and solar cells, especially those in which the sun's energy is transformed into electrical or chemical. From the ecology point of view, using solar energy does not disturb the thermal balance of our planet, either being directly converted into heat in solar collectors or being transformed into electrical or chemical energy in solar cells and batteries. On the other hand, every kilowatt hour of energy thus obtained replaces a certain amount of fossil or nuclear fuel and mitigates any associated adverse effects known. Solar energy is considered to be one of the most sustainable energy resources for future energy supplies.

To make the energy of solar radiation converted into electricity, materials that behave as semiconductors are used. Semiconductive properties of copper sulfides and copper oxides, as well as compounds of chalcopyrite type have been extensively investigated (Rajčić-Vujasinović et al., 1994, 1999). One of the important design criteria in the development of an effective solar cell is to maximize its efficiency in converting sunlight to electricity. A photovoltaic cell consists of a light absorbing material which is connected to an external circuit in an asymmetric manner. Charge carriers are generated in the material by the absorption of photons of light, and are driven towards one or other of the contacts by the built-in spatial asymmetry. This light driven charge separation establishes a photo voltage at open circuit, and generates a photocurrent at short circuit. When a load is connected to the external circuit, the cell produces both current and voltage and can do electrical work.

Solar technology, thanks to its advantages regarding the preservation of the planetary energy balance, is getting into an increasing number of application areas. So, for example, Rizzo et al. (2010) as well as Stević & Rajčić-Vujasinović (in Press) describe hybrid solar vehicles, while Vieira & Mota (2010) show a rechargeable battery with photovoltaic panels.

The high cost of silicon solar cells forces the development of new photovoltaic devices utilizing cheap and non-toxic materials prepared by energy-efficient processes. The Cu–O system has two stable oxides: cupric oxide (CuO) and cuprous oxide (Cu_2O). These two oxides are semiconductors with band gaps in the visible or near infrared regions. Copper and copper oxide (metal-semiconductor) are one of the first photovoltaic cells invented (Pollack and Trivich, 1975). Cuprous oxide (Cu_2O) is an attractive semiconductor material that could be used as anode material in thin film lithium batteries (Lee et al, 2004) as well as in solar cells (Akimoto et al., 2006; Musa et al., 1998; Nozik et al., 1978; Tang et al., 2005). Its semiconductor properties and the emergence of photovoltaic effect were discovered by Edmond Becquerel 1839th[1] experimenting in the laboratory of his father, Antoine-César Becquerel.

Cu_2O is a p-type semiconductor with a direct band gap of 2.0–2.2 eV (Grozdanov, 1994) which is suitable for photovoltaic conversion. Tang et al. (2005) found that the band gap of nanocrystalline Cu_2O thin films is 2.06 eV, while Siripala et al. (1996) found that the deposited cuprous oxide exhibits a direct band gap of 2.0 eV, and shows an n-type behavior when used in a liquid/solid junction. Han & Tao (2009) found that n-type Cu_2O deposited in a solution containing 0.01 M copper acetate and 0.1 M sodium acetate exhibits higher resistivity than p-type Cu_2O deposited at pH 13 by two orders of magnitude. Other authors, like Singh et al. (2008) estimated the band gap of prepared Cu_2O nanothreads and nanowires to be 2.61 and 2.69 eV, which is larger than the direct band gap (2.17 eV) of bulk Cu_2O (Wong & Searson, 1999). The higher band gap can be attributed to size effect of the present nanostructures. Thus the increase of band gap as compared to the bulk can be understood on the basis of quantum size effect which arises due to very small size of nanothreads and nanowires in one-dimension.

Cuprous oxide attracts the most interest because of its high optical absorption coefficient in the visible range and its reasonably good electrical properties (Musa et al., 1998). Its advantages are, in fact, relatively low cost and low toxicity. Except for a thin film that can be electrochemically formed on different substrates (steel, TiO_2), cuprous oxide can be obtained in the form of nano particles with all the benefits offered by nano-technology (Daltin et al., 2005; Zhou & Switzer, 1998). Nanomaterials exhibit novel physical properties and play an important role in fundamental research.

The unit cell of Cu_2O with a lattice constant of 0.427 nm is composed of a body centered cubic lattice of oxygen ions, in which each oxygen ion occupies the center of a tetrahedron formed by copper ions (Xue & Dieckmann, 1990). The Cu atoms arrange in a fcc sublattice, the O atoms in a bcc sublattice. The unit cell contains 4 Cu atoms and 2 O atoms. One sublattice is shifted by a quarter of the body diagonal. The space group is Pn3m, which includes the point group with full octahedral symmetry. This means particularly that parity is a good quantum number. Figure 1 shows the crystal lattice of Cu_2O. Molar mass of Cu_2O is 143.09 g/mol, density is 6.0 g/cm^3 and its melting and boiling points are 1235°C and 1800°C, respectively. Also, it is soluble in acid and insoluble in water.

Cuprous oxide (copper (I) oxide Cu_2O) is found in nature as cuprite and formed on copper by heat. It is a red color crystal used as a pigment and fungicide. Rectifier diodes based on this material have been used industrially as early as 1924, long before silicon became the standard. Cupric oxide (copper(II) oxide CuO) is a black crystal. It is used in making fibers and ceramics, gas analyses and for Welding fluxes. The biological property of copper compounds takes important role as fungicides in agriculture and biocides in antifouling paints for ships and wood preservations as an alternative of Tributyltin compounds.

In solar cells, Cu_2O has not been commonly used because of its low energy conversion efficiency which results from the fact that the light generated charge carriers in micron-sized Cu_2O grains are not efficiently transferred to the surface and lost due to recombination. For randomly generated charge carriers, the average diffusion time from the bulk to the surface is given by:

---

[1] http://pvcdrom.pveducation.org/MANUFACT/FIRST.HTM

$$\tau = r^2 / \pi^2 D \qquad (1)$$

where r is the grain radius and D is the diffusion coefficient of the carrier (Rothenberger et al., 1985, as cited in Tang et al., 2005). If the grains radius is reduced from micrometer dimensions to nanometer dimensions, the opportunities for recombination can be dramatically reduced. The preparation of nano crystalline Cu_2O thin films is a key to improving the performance of solar application devices. Nanotechnologies in this area, therefore, given their full meaning. In the last decade the scientific literature, abounds with works again showing progress in research related to obtaining the cuprous oxide.

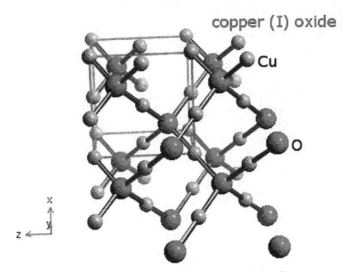

(http://www.webelements.com/compounds/copper/dicopper_oxide.html)

Fig. 1. Crystal structure of Cu_2O

This chapter presents an overview of recent literature concerning cuprous oxide synthesis and application as an active material in solar cells, as well as our own results of synthesis and investigations of Cu_2O thin films using electrochemical techniques.

2. Methodologies used for the synthesis of cuprous oxide

The optical and electrical properties of absorber materials in solar cells are key parameters which determine the performance of solar cells. Hence, it is necessary to tune these properties properly for high efficient device. Electrical properties of Cu_2O, such as carrier mobility, carrier concentration, and resistivity are very dependent on preparation methods. Cuprous oxide thin films have been prepared by various techniques like thermal oxidation (Jayatissa et al., 2009; Musa et al., 1998; Sears & Fortin, 1984), chemical vapor deposition (Kobayashi et al. 2007; Maruyama, 1998; Medina-Valtierra et al., 2002; Ottosson et al., 1995; Ottosson & Carlsson, 1996), anodic oxidation (Fortin & Masson, 1982; Sears and Fortin, 1984; Singh et al., 2008), reactive sputtering (Ghosh et al., 2000), electrodeposition (Briskman, 1992; Daltin et al., 2005; Georgieva & Ristov, 2002; Golden et al., 1996; Liu et al., 2005; Mizuno et al., 2005; Rakhshani et al., 1987, Rakhshani & Varghese, 1987; Santra et al., 1999; Siripala et al., 1996; Tang et al., 2005; Wang et al., 2007; Wijesundera et al., 2006), plasma evaporation (Santra et al., 1992), sol–gel-like dip technique (Armelao et al., 2003; Ray, 2001) etc. Each of these methods has its own advantages and disadvantages. In most of these studies, a mixture of phases of Cu, CuO and Cu_2O is generally obtained and this is one of the nagging problems for non-utilizing Cu_2O as a semiconductor (Papadimitropoulos et al., 2005). Pure Cu_2O films can be obtained by oxidation of copper layers within a range of temperatures followed by annealing for a small period of time.

Results obtained using different methods, especially thermal oxidation and chemical vapor evaporation for synthesis of cuprous oxide thin films, are presented in next sections, with special emphasis on the electrochemical synthesis of cuprous oxide.

2.1 Thermal oxidation

Polycrystalline cuprous oxide can be formed by thermal oxidation of copper under suitable conditions (Rai, 1988). The procedure involves the oxidation of high purity copper at an elevated temperature (1000–1500^{0}C) for times ranging from few hours to few minutes depending on the thickness of the starting material (for total oxidation) and the desired thickness of Cu_2O (for partial oxidation). Process is followed by high-temperature annealing for hours or even days.

Sears & Fortin (1984) synthesized cuprous oxide films on copper substrates to a thickness of a few micrometers, using both thermal and anodic oxidation techniques. The measurements carried out on the anodic oxide layers indicate an unwanted but inevitable incorporation of other compounds into the Cu_2O. They found that the photovoltaic properties of the resulting Cu_2O/Cu backwall cells depend critically on the copper surface preparation, as well as on the specific conditions of oxidation. Backwall cells of the thermal variety with thicknesses down to 3 μm do not quite yet approach the performance of the best Cu_2O front cells, but are much simpler to grow. Serious difficulties with shorting paths in the case of thermally grown oxide and with the purity of the Cu_2O in the anodic case will have to be solved before a solar cell with an oxide layer thickness in the 1.5 to 2 μm range can be produced.

Musa et al. (1998) produced the cuprous oxide by thermal oxidation and studied its physical and electrical properties. The oxidation was carried out at atmospheric pressure in a high-temperature tube furnace. During this process the copper foils were heated in the range of 200 to 1050°C. Cu_2O has been identified to be stable at limited ranges of temperature and oxygen pressure. It has also been indicated that during oxidation, Cu_2O is formed first, and after a sufficiently long oxidation time CuO is formed (Roos & Karlson, 1983, as cited Musa et al., 1998). It has been suggested that the probable reactions that could account for the presence of CuO in layers oxidised below 1000 °C are:

$$2Cu_2O + O_2 \rightarrow 4CuO \qquad (2)$$

$$Cu_2O \rightarrow CuO + Cu \qquad (3)$$

The unwanted CuO can be removed using an etching solution consisting of FeCl, HCl, and 8 M HNO_3 containing NaCl. The results of the oxidation process as deduced from both XRD and SEM studies indicate that the oxide layers resulting from oxidation at 1050^{0}C consist entirely of Cu_2O. Those grown below 1040^{0}C gave mixed oxides of Cu_2O and CuO. It was observed that in general the lower the temperature of oxidation, the lower the amount of Cu_2O was present in the oxide. Thermodynamic considerations indicate that the limiting temperature for the elimination of CuO from the oxide layer was found to be 1040°C. For thermal oxidation carried out below 1040°C, Cu_2O is formed first and it is then gradually oxidised to CuO depending on the temperature and time of reaction. Pure unannealed Cu_2O layers grown thermally in air are observed to exhibit higher resistivity and low hole mobility. A significant reduction in resistivity and an increase in mobility values were obtained by oxidizing the samples in the presence of HCl vapour, followed by annealing at 500°C. Cu_2O layers grown in air without the annealing process gave resistivities in the range $2x10^3 - 3x10^3$ Ωcm. A substantial reduction in the resistivity of the samples was achieved by doping with chlorine during growth and annealing. An average mobility of 75 cm^2 V^{-1} s^{-1} at room temperature was obtained for eight unannealed Cu_2O samples. This average value increased to 130 cm^2 V^{-1} s^{-1} after doping the samples with chlorine and annealing. The SEM studies indicate that the annealing process results in dense polycrystalline Cu_2O layers of increased grain sizes which are appropriate for solar-cell fabrication. Figure 2 presents the micrograph of the surface morphology of a copper foil partially oxidised at 970°C for 2 min. The sample was neither annealed nor etched. The surface shows the black CuO coat formed on the violet-red Cu_2O after the oxidation process. The surface morphology is porous and amorphous in nature. The structure formed by this oxidation process is of the form $CuO/Cu_2O/Cu/Cu_2O/CuO$.

Jayatissa et al. (2009) prepared cuprous oxide (Cu_2O) and cupric oxide (CuO) thin films by thermal oxidation of copper films coated on indium tin oxide (ITO) glass and non-alkaline glass substrates. The formation of Cu_2O and CuO was controlled by varying oxidation conditions such as oxygen partial pressure, heat treatment temperature and oxidation time. Authors used X-ray diffraction, atomic force microscopy and optical spectroscopy to determinate the microstructure, crystal direction, and optical properties of copper oxide films. The experimental results suggest that the thermal oxidation method can be employed to fabricate device quality Cu_2O and CuO films that are up to 200–300 nm thick.

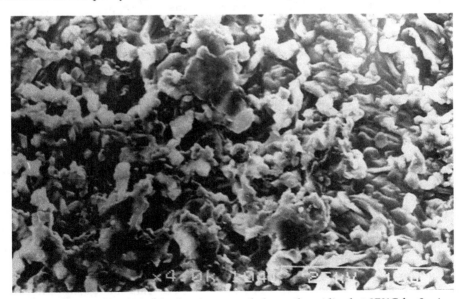

Fig. 2. SEM micrograph of unetched and unannealed sample oxidised at 970°C for 2 min showing CuO coating (Musa et al., 1998)

2.2 Chemical vapor deposition

Chemical vapor deposition is a chemical process used to produce high-purity, high-performance solid materials. The films may be epitaxial, polycrystalline or amorphous depending on the materials and reactor conditions. Chemical vapor deposition has become the major method of film deposition for the semiconductor industry due to its high throughput, high purity, and low cost of operation. Several important factors affect the quality of the film deposited by chemical vapor deposition such as the deposition temperature, the properties of the precursor, the process pressure, the substrate, the carrier gas flow rate and the chamber geometry.

Maruyama (1998) prepared polycrystalline copper oxide thin films at a reaction temperature above $280^{0}C$ by an atmospheric-pressure chemical vapor deposition method. Copper oxide films were grown by thermal decomposition of the source material with simultaneous reaction with oxygen. At a reaction temperature above $280^{0}C$, polycrystalline copper oxide films were formed on the borosilicate glass substrates. Two kinds of films, i.e., Cu_2O and CuO, were obtained by adjusting the oxygen partial pressure. Also, there are large differences in color and surface morphology between the CuO and Cu_2O films obtained. Author found that the surface morphology and the color of CuO film change with reaction temperature. The CuO film prepared at $300^{0}C$ is real black, and the film prepared at $500^{0}C$ is grayish black.

Medina-Valtierra et al. (2002) coated fiber glass with copper oxides, particularly in the form of $6CuO \bullet Cu_2O$ by chemical vapor deposition method. The authors' work is based on design of an experimental procedure for obtaining different copper phases on commercial fiberglass. Films composed of copper oxides were deposited over fiberglass by sublimation and transportation of $(acac)_2Cu(II)$ with a O_2 flow (oxidizing agent), resulting in the decomposition of the copper precursor, deposition of Cu^0 and Cu^0 oxidation on the fiberglass over a short range of deposition temperatures. The copper oxide films on the fiberglass were examined using several techniques such as X-ray diffraction (XRD), visible spectrophotometry, scanning electronic microscopy (SEM) and atomic force microscopy (AFM). The films formed on fiberglass showed three different colors: light brown, dark brown and gray when Cu_2O, $6CuO \bullet Cu_2O$ or CuO, respectively, were present. At a temperature of $320°C$ only cuprous oxide is formed but at a higher temperature of about $340°C$ cupric oxide is formed. At a temperature of $325°C$ $6CuO-Cu_2O$ is formed. The decomposition of precursor results in the formation of a zero valent copper which upon oxidation at different temperature gives different oxides.

Ottosson et al. (1995) deposited thin films of Cu_2O onto MgO (100) substrates by chemical vapour deposition from copper iodide (CuI) and dinitrogen oxide (N_2O) at two deposition temperatures, $650°C$ and $700°C$. They found that the pre-treatment of the substrate as well as the deposition temperature had a strong influence on the orientation of the nuclei and the film. For films deposited at $650°C$ several epitaxial orientations were observed: (100), (110) and (111). The Cu_2O(100) was found to grow on a defect MgO(100) surface. When the substrates were annealed at $800°C$ in N_2O for 1 h, the defects in the surface disappeared and only the (110) orientation was developed during the deposition. The films deposited at $700°C$ (without annealing of the substrates) displayed only the (110) orientation.

Markworth et al. (2001) prepared cuprous oxide (Cu_2O) films on single-crystal MgO(110) substrates by a chemical vapor deposition process in the temperature range $690–790°C$. Cu_2O ($a=0.4270$ nm) and MgO ($a=0.4213$ nm) have cubic crystal structures, and the lattice mismatch between them is 1.4%. Due to good lattice match, chemical stability, and low cost,

MgO single crystals are particularly effective substrates for the growth of Cu_2O thin films. Authors found that the Cu_2O films grow by an island-formation mechanism on MgO substrate. Films grown at 690°C uniformly coat the substrate except for micropores between grains. However, at a growth temperature of 790°C, an isolated, three-dimensional island morphology develops.

Kobayashi et al. (2007) investigated the high-quality Cu_2O thin films grown epitaxially on MgO (110) substrate by halide chemical vapor deposition under atmospheric pressure. CuI in a source boat was evaporated at a temperature of 883 K, and supplied to the growth zone of the reactor by N_2 carrier gas, and O_2 was also supplied there by the same carrier gas. Partial pressure of CuI and O_2 were adjusted independently to 1.24×10^{-2} and 1.25×10^3 Pa. They found that the optical band gap energy of Cu_2O film calculated from absorption spectra is 2.38 eV. The reaction of CuI and O_2 under atmospheric pressure yields high-quality Cu_2O films.

2.3 Other methods

Several novel methods for the synthesis of cuprous oxide (i.e. reactive sputtering, sol-gel technique, plasma evaporation,) and some results obtained using these techniques are presented in this part. For example, Santra et al. (1992) deposited thin films of cuprous oxide on the substrates by evaporating metallic copper through a plasma discharge in the presence of a constant oxygen pressure. Authors found two oxide phases before and after annealing treatment of films. Before annealing treatment, cuprous oxide was identified and after annealing in a nitrogen atmosphere, cuprous oxide changes to cupric oxide. The results of optical absorption measurement show that the band gap energies for Cu_2O and CuO are 2.1 eV and 1.85 eV, respectively. Thin films prepared in the absence of a reactive gas and plasma were also deposited on glass substrates and in these films the presence of metallic copper was identified.

Ghosh et al. (2000) deposited cuprous oxide and cupric oxide by RF reactive sputtering at different substrate temperatures, namely, at 30, 150 and 300ºC. They used atomic force microscopy for examination of the properties of the prepared oxides films related to surface morphology. It was found for the film deposited at 30ºC, that, 8-10 small grains of size ~40 nm diameter agglomerate together and make a big grain of size ~120 nm. At the temperature of 150ºC the grain size becomes 160 nm. The grain size decreases to 90 nm at 300ºC. From thickness and deposition time, the deposition rates of the films are found to be 8, 11.5 and 14.0 nm/min for substrate temperature corresponding to 30, 150 and 300ºC, respectively. Optical band gap of the films deposited at 30, 150 and 300ºC are 1.75, 2.04 and 1.47 eV, respectively. Different phases of copper oxides are found at different temperatures of deposition. CuO phase is obtained in the films prepared at a substrate temperature of 300ºC.

Sol gel-like dip technique is a very simple and low-cost method, which requires no sophisticated specialized setup. For example, Armelao et al. (2003) used a sol-gel method to synthesize nanophasic copper oxide thin films on silica slides. They used copper acetate monohydrate as a precursor in ethanol as a solvent. Authors observed formation of CuO crystallites in the samples annealed under inert atmosphere (N_2) up to 3 h. A prolonged treatment (5 h) in the same environment resulted in the complete disappearance of tenorite and in the formation of a pure cuprite crystalline phase. Also, under reducing conditions, the formation of CuO, Cu_2O and Cu was progressively observed, leading to a mixture of Cu(II) and Cu(I) oxides and metallic copper after treatment at 900ºC for 5 h.

All the obtained films have nanostructure with an average crystallite size lower than 20 nm.

Nair et al. (1999) deposited cuprous oxide thin films on glass substrate using chemical technique. The glass slides were dipped first in a 1 M aqueous solution of NaOH at the temperature range 50-90°C for 20 s and then in a 1 M aqueous solution of copper complex. X-ray diffraction patterns showed that the films, as prepared, are of cuprite structure with composition Cu_2O. Annealing the films in air at 350°C converts these films to CuO. This conversion is accompanied by a shift in the optical band gap from 2.1 eV (direct) to 1.75 eV (direct). The films show p-type conductivity, $\sim 5 \times 10^{-4}$ Ω^{-1} cm^{-1} for a film of thickness 0.15 μm.

3. Electrochemical synthesis

3.1 Electrodeposition

Synthesis of Cu_2O nanostructures by the methods described in the previous part demands complex process control, high reaction temperatures, long reaction times, expensive chemicals and specific method for specific nanostructures. A request for obtaining nanometer particles, cause complete change of technology in which Cu_2O is formed on the cathode by reduction of Cu^{2+} ions from the organic electrolyte. The possible reactions during the cathodic reduction of copper (II) lactate solution are:

$$2Cu^{2+} + H_2O + 2e^- = Cu_2O + 2H^+ \qquad (4)$$

$$Cu^{2+} + 2e^- = Cu \qquad (5)$$

$$Cu_2O + 2H^+ + 2e^- = 2Cu + H_2O \qquad (6)$$

The electrodeposition techniques are particularly well suited for the deposition of single elements but it is also possible to carry out simultaneous depositions of several elements and syntheses of well-defined alternating layers of metals and oxides with thicknesses down to a few nm. So, electrodeposition is a suitable method for the synthesis of semiconductor thin films such as oxides. This method provides a simple way to deposit thin Cu(I) oxide films onto large-area conducting substrates (Lincot, 2005). Thus, the study of the growth kinetics of these films is of considerable importance. In this section we present some results of electrochemical deposition of cuprous oxide obtained by various authors.

Rakhshani et al. (1987) cathodically electrodeposited Cu(I) oxide film onto conductive substrates from a solution of cupric sulphate, sodium hydroxide and lactic acid. Films of Cu_2O were deposited in three different modes, namely the potentiostatic mode, the mode with constant WE potential with respect to the CE and the galvanostatic mode. The composition of the films deposited under all conditions was Cu_2O with no traces of CuO. The optical band gap for electrodeposited Cu_2O films was 1.95 eV. Deposition temperature played an important role in the size of deposited grains. Films were photoconductive with high dark resistivities. Also, Rakhshani & Varghese (1987) electrodeposited cuprous oxide thin films galvanostatically on 0.05 mm thick stainless steel substrates at a temperature of 60°C. The deposition solution with pH 9 consisted of lactic acid (2.7 M), anhydrous cupric sulphate (0.4 M), and sodium hydroxide (4 M). Authors found that all the films deposited at 60 °C consisted only of Cu_2O grains a few μm in size and preferentially oriented along (100) planes parallel to the substrate surface. A band gap was found and it was 1.90-1.95 eV.

Mukhopadhyay et al. (1992) deposited Cu_2O films by galvanostatic method on copper substrates. An alkaline cupric sulphate (about 0.3 M) bath containing NaOH (about 3.2 M) and lactic acid (about 2.3 M) was used as the electrolyte at pH 9. The bath temperatures were 40, 50 and 60°C. XRD analysis indicated a preferred (200) orientation of the Cu_2O deposited film. The deposition kinetics was found to be independent of deposition temperature and linear in the thickness range studied (up to about 20 μm). The electrical conductivity of Cu_2O films was found to vary exponentially with temperature in the 145-300°C range with associated activation energy of 0.79 eV.

Golden et al. (1996) found that the reflectance and transmittance of the electrodeposited films of cuprous oxide give a direct band gap of 2.1 eV. Namely, authors used electrodeposition method for obtaining the films of cuprous oxide by reduction of copper (II) lactate in alkaline solution (0.4 M cupric sulfate and 3 M lactic acid). Films were deposited onto either stainless steel or indium tin oxide (ITO) substrates. Deposition temperatures ranged from 25 to 65 °C. They found that the cathodic deposition current was limited by a Schottky-like barrier that forms between the Cu_2O and the deposition solution. A barrier height of 0.6 eV was determined from the exponential dependence of the deposition current on the solution temperature. At a solution pH 9 the orientation of the film is [100], while at a solution pH 12 the orientation changes to [111]. The degree of [111] texture for the films grown at pH 12 increased with applied current density.

Siripala et al. (1996) deposited cuprous oxide films on indium tin oxide (ITO) coated glass substrates in a solution of 0.1 M sodium acetate and 1.6 x 10^{-2} M cupric acetate and the effect of annealing in air has been studied too. Electrodeposition was carried out for 1.5 h in order to obtain films of thicknesses in the order of 1 μm. Authors concluded that the electrodeposited Cu_2O films are polycrystalline with grain sizes in the order of 1-2 μm and the bulk crystal structure is simple cubic. They concluded that there is no apparent change in the crystal structure when heat treated in air at or below 300°C. Annealing in air changes the morphology of the surface creating a porous nature with ring shaped structures on the surface. Annealing above 300°C causes decomposition of the yellow-orange colour Cu_2O film into a darker film containing black CuO and its complexes with water.

Zhou & Switzer (1998) deposited Cu_2O films on stainless steel disks by the cathodic reduction of copper (II) lactate solution (0.4 M cupric sulfate and 3 M lactic acid). The pH of the bath was between 7 and 12 and the bath temperature was 60°C. Authors concluded that the preferred orientation and crystal shape of Cu_2O films change with the bath pH and the applied potential. They obtained pure Cu_2O films at bath pH 9 with applied potential between -0.35 and -0.55 (SCE) or at bath pH 12.

Mahalingam et al. (2000) deposited cuprous oxide thin films on copper and tin-oxide-coated glass substrates by cathodic reduction of alkaline cupric lactate solution (0.45 M $CuSO_4$, 3.25 M lactic acid and 0.1 M NaOH). The deposition was carried out in the temperature range of 60-80°C at pH 9. Galvanostatic deposition on tin-oxide-coated glass and copper substrates yields reddish-grey Cu_2O films. All the films deposited are found to be polycrystalline having grains in the range of 0.01 - 0.04 μm. The deposition kinetics is found to be linear and independent of the deposition temperature. From the optical absorption measurements, authors found that the deposit of cuprous oxide films has a refractive index of 2.73, direct band gap of 1.99 eV, and extinction coefficient of 0.195. After deposition on temperature of 70°C, cuprous oxide films were annealed in air for 30 min at different temperatures (150, 250 and 350ºC) to obtain their room temperature resistivity. It showed a decrease in resistivity of Cu_2O film of the order of 10^7 Ωcm to 10^4 Ωcm. The explanation of such behavior may be due to increase in hole conduction.

Georgieva & Ristov (2002) deposited the cuprous oxide (Cu_2O) films using a galvanostatic method from an alkaline $CuSO_4$ bath containing lactic acid and sodium hydroxide (64 g/l anhydrous cupric sulphate ($CuSO_4$), 200 ml/l lactic acid ($C_3H_6O_3$) and about 125 g/l sodium hydroxide (NaOH)). The electrodeposition temperature was 60ºC. Authors obtained polycrystalline films of 4–6 μm in thickness with optical band gap of 2.38 eV.

Daltin et al. (2005) applied potentiostatic deposition method to obtain cuprous oxide nanowires in polycarbonate membrane by cathodic reduction of alkaline cupric lactate solution (0.45 M Cu(II) and 3.25 M lactate). Authors found that the optimum electrochemical parameters for the deposition of nanowires are: pH 9.1, temperature 70ºC, and applied potential -0.9 V (SSE). The morphology of the nanowires was analyzed by SEM. The obtained nanowires had uniform diameters of about 100 nm and lengths up to 16 μm. Scanning electron micrograph of electrodeposited Cu_2O nanowires are presented in Figure 3.

Liu et al. (2005) investigated the electrochemical deposition of Cu_2O films onto three different substrates (indium tin oxide film coated glass, n-Si wafer with (001) orientation and Au film evaporated onto Si substrate). For the film grown on ITO, electrical current increases gradually during deposition, while for the films growth on both Si and Au substrates, the monitored current decreases monotonically. Authors considered that the continuous decrease in current reflects different deposition mechanisms. In the case of Si substrate, the decrease of the current may be the result of the formation of an amorphous SiO_2 layer on the Si surface, which limits the current. For the Au surface, the decrease in measured current is due to the resistivity increase as a result of Cu_2O film formation. Cu_2O crystals with microsized pyramidal shape were grown on ITO substrate. Nanosized and pyramidal shaped Cu_2O particles were formed on Si substrate and the film grown on Au substrate shows a (100) orientation with much better crystallinity.

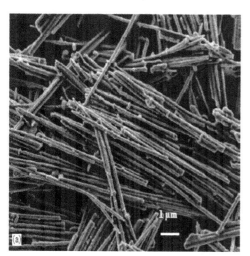 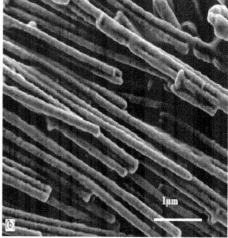

Fig. 3. (a) Scanning electron micrograph of electrodeposited Cu_2O nanowires. Bath temperature 70ºC, pH 9.1, E -1.69 V/$_{SSE}$. (b) Enlarged (a) (Daltin et al., 2005)

Tang et al. (2005) investigated the electrochemical deposition of nanocrystalline Cu_2O thin films on TiO_2 films coated on transparent conducting optically (TCO) glass substrates by cathodic reduction of cupric acetate (0.1 M sodium acetate and 0.02 M cupric acetate). Authors concluded that the pH and bath temperature strongly affect the composition and microstructure of the Cu_2O thin films. The effect of bath pH on electrodeposition of Cu_2O thin film was investigated by selecting a bath temperature of 30°C and an applied potential of -245 mV (SCE). Authors found that the films deposited at pH 4 are mostly metallic Cu and only little Cu_2O. In the region of pH 4 to pH 5.5, the deposited films are a composite of Cu and Cu_2O, while the films deposited at pH between 5.5 and 6 are pure Cu_2O. Pure Cu_2O deposited at bath temperature between 0 and 30°C produced spherically shaped grains with 40~50 nm in diameter. The bath temperature must be controlled in the range of 0-30°C to obtain nanocrystalline Cu_2O thin film. At a temperature of 45°C, a highly branched dendrite formed, and the grain size increased to 200–500 nm. At the temperature above 60°C, a ring-shaped structure with a porous surface was observed. Optical absorption measurements indicate that annealing at 200°C can improve the transmittance of the nanocrystalline Cu_2O thin films. Figure 4 shows SEM photographs of Cu_2O films deposited at various bath temperatures.

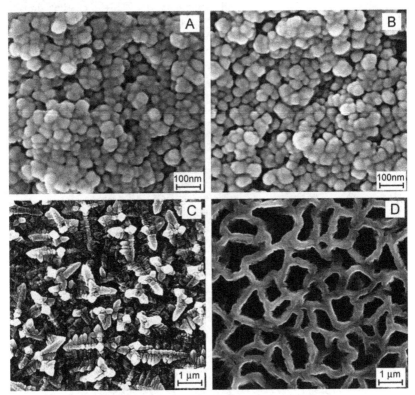

Fig. 4. SEM photographs of Cu_2O films deposited at various bath temperatures: (A) 0°C, (B) 30°C, (C) 45°C, and (D) 60°C (Tang et al., 2005)

Wijesundera et al. (2006) investigated the potentiostatic electrodeposition of cuprous oxide and copper thin films. Electrodeposition was carried out in an aqueous solution containing sodium acetate and cupric acetate. The results of their investigation show that the single phase polycrystalline Cu_2O can be deposited from 0 to -300 mV (SCE). Also, co-deposition of Cu and Cu_2O starts at - 400 mV (SCE). At the deposition potential from -700 mV (SCE) a single phase Cu thin films are produced. Single phase polycrystalline Cu_2O thin films with cubic grains of 1–2 µm can be possible at the deposition potential around -200 mV (SCE).

Wang et al. (2007) cathodically electrodeposited cuprous oxide films from 0.4 M copper sulfate bath containing 3 M lactic acid. The bath pH was carefully adjusted between 7.5 and 12.0 by controlled addition of 4 M NaOH. The electrodeposition was done on Sn-doped indium oxide substrates. The influence of electrodeposition bath pH on grain orientation and crystallite shape was examined. Authors found that three orientations, namely, (100), (110), and (111) dominate as the bath pH is increased from ~ 7.5 to ~ 12.

Recently, Hu et al. (2009) electrodeposited Cu_2O thin films onto an indium tin oxide (ITO) coated glass by a two-electrode system with acid and alkaline electrolytes under different values of direct current densities. Copper foils were used as the anodes, and the current density between the anode and cathode varied between 1 mA cm^{-2} and 5 mA cm^{-2}. It was obtained that the microstructure of Cu_2O thin films produced in the acid electrolyte changes from a ring shape to a cubic shape with the increase of direct current densities. The microstructure of Cu_2O thin films produced in the alkaline electrolyte has a typical pyramid shape. The electrocrystallization mechanisms are considered to be related to the nucleation rate, cluster growth, and crystal growth. To investigate the initial stage of nucleation and cluster growth, different current densities with the same deposition time were applied. Figure 5 shows that a relatively large cluster size and a relatively small number of nucleation sites were obtained under a current density of 1 mAcm^{-2}. At a high current density of 5 mAcm^{-2}, more nucleation sites and a small cluster size were obtained.

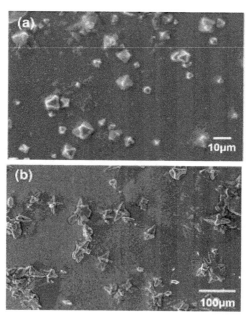

Fig. 5. The Cu_2O films synthesized under different current densities with the same deposition time (Hu et al., 2009)

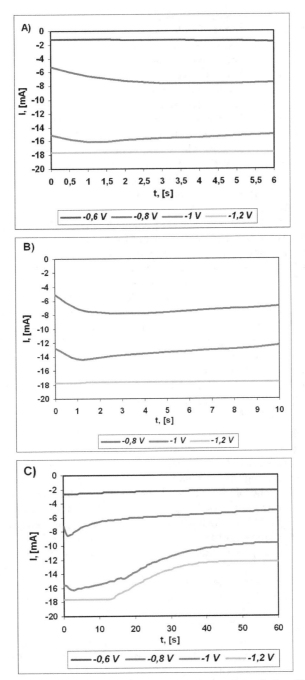

Fig. 6. Current density vs. time curves for electrodeposition of Cu₂O thin film on titanium electrode (electrodeposition time: (A) 6 s, (B) 10 s and (C) 60 s; t = 25 °C, pH 9.22)

Bugarinović et al. (2009) investigated the electrochemical deposition of thin films of cuprous oxide on three different substrates (stainless steel, platinum and copper). All experiments of Cu_2O thin films deposition were performed at room temperature. Using experimental technique described elsewhere (Stević & Rajčić-Vujasinović, 2006; Stević & et al., 2009), electrodeposition was carried out in in a copper lactate solution as an organic electrolyte (0.4 M copper sulfate and 3 M lactic acid, pH 7-10 is set using NaOH). The conditions are adjusted so that the potentials which arise Cu_2O and CuO are as different as possible. Characterization of obtained coatings was performed by cyclic voltammetry. The results indicate that the composition of the substrate strongly affects electrochemical reactions. Reaction with the highest rate took place on a copper surface, while the lowest rate was obtained on the platinum electrode. The results show that the co-deposit of Cu_2O and Cu was obtained at - 800 mV (SCE) on stainless steel electrode. The same authors investigated the electrodeposition of cuprous oxide thin film on titanium electrode. The obtained results are presented in Figure 6.

Cuprous oxide thin films were deposited at potentials -0.6 V, -0.8 V, -1.0 V and 1.2 V with respect to SCE. All experiments were carried out for a duration of 6 s, 10 s and 1 minute. When the electrodeposition lasted 6 s (Fig. 6A), obtained currents depended on applied potentials. Lowest current of 1mA was obtained at the potential of -0.6 V vs. SCE, while the highest value of 17.9 mA was reached at -1.2 V (SCE). When the electrodeposition time was 10 s (Fig. 6B), curves current vs. time had similar shape as the previous, but when the process duration prolongates to 60 s (Fig. 6C), currents obtained at higher potentials (-1.0 V and -1.2 V vs. SCE) decrease after about 15 s and stabilise again after about 40 s at some lower value (nearly 80% of the previous ones). Maximum theoretical thicknesses of Cu_2O film for every applied potential and all process durations were calculated. The lowest thickness of 7 nm was obtained for 6 s with potential of -0.6 V (SCE). More negative potentials and the increase of time lead to the increase of the film thickness. Theoretical value of the Cu_2O film thickness for the longest time (60 s) and most negative potential (-1.2 V vs. SCE) is about 900 nm.

3.2 Anodic oxidation

In spite of the simple equipment and easy process control, cathodic synthesis demands expensive chemicals as a big dissadventage. On the other hand, anodic oxidation of copper in alkaline solution is one of the standard methodologies for producing cuprous oxide powders used for marine paints and for plants preservation. Those powders are composed of particles of micrometer scale. However, solar sells, for their part, require particles or films of much smaller dimensions in order to achieve higher efficiency. Passive protecting layers formed on copper during anodic oxidation in alkaline solutions are widely investigated and described in electrochemical literature. The structure of those films formed on copper in neutral and alkaline solutions consists mainly of Cu_2O and CuO or $Cu(OH)_2$. Applying in situ electrochemical scanning tunneling microscopy (STM), Kunze et al. (2003) found that in NaOH solutions, a Cu_2O layer is formed at E > 0.58-0.059 pH (V vs. SHE). A $Cu_2O/Cu(OH)_2$ duplex film is found for E > 0.78-0.059 pH (V vs. SHE). In borate buffer solutions, oxidation to Cu_2O leads to non-crystalline grain like structure, while a crystalline and epitaxial Cu_2O layer has been observed in 0.1 M NaOH indicating a strong anion and/or pH effect on the crystallinity of the anodic oxide film.

Stanković et al. (1998; 1999) investigated the effect of different parameters such as temperature, pH and anodic current density on CuO powder preparation. The lowest value of average crystallite size was obtained at pH 7.5, whereas the highest value was obtained at pH 9.62. They found a strong dependence of grain size and cupric oxide purity on current density. The average srystallite size increased from 45 nm (at a current density of 500 Am^{-2}) to 400 nm (at a current density of 4000 Am^{-2}), other conditions being as follows: pH 7.5, temperature of 353 K and 1.5 M Na_2SO_4.

There have been a number of papers on anodic formation of thin Cu_2O layers (< 1 μm) using alkaline solutions, but some work has been done with slightly acidic solutions. For example, backwall Cu_2O/Cu photovoltaic cells have been prepared by Sears and Fortin (Sears & Fortin, 1983) with the Cu_2O layer being about 1 μm thick. They used and compared two methods of oxidation – thermal and anodic. The condition of the underlying copper surface is expected to influence the resulting parameters of thin solar cells, so they examined the influence of the surface preparation of the starting copper (i.e., polishing technique, thermal annealing). All this experience can help in researching the optimal way of production of nanostructured Cu_2O powders or films.

Recently, Singh et al. (2008) reported synthesis of nanostructured Cu_2O by anodic oxidation of copper through a simple electrolysis process employing plain water as electrolyte. They found two different types of Cu_2O nanostructures. One of them belonged to particles collected from the bottom of the electrolytic cell, while the other type was located on the copper anode itself. The Cu_2O structures collected from the bottom consist of nanowires (length, ~ 600–1000 nm and diameter, ~ 10–25 nm). It may be mentioned that the total length of Cu_2O nanothread and nanowire is comprised of several segments. These were presumably formed due to interaction between nanothreads/nanowires forming the network in which the Cu_2O nanothread/nanowire configuration finally appears. When the electrolysis conditions were maintained at 10 V for 1 h, the representative TEM microstructure revealed the presence of dense Cu_2O nanowire network (length, ~ 1000 nm, diameter, ~ 10–25 nm). The X-ray diffraction pattern obtained from these nanomaterials, could be indexed to a cubic system with lattice parameter, a = 0.4269 ± 0.005 nm. These tally quite well with the lattice parameter of Cu_2O showing that the material formed under electrolysis conditions consists of cubic Cu_2O lattice structure.

In addition to the delaminated nanostructures, investigations of the copper anode, which were subjected to electrolysis runs, revealed the presence of another type of nanostructure of Cu_2O. Authors propose that the higher applied voltage (e.g. 8 V or 10 V) for electrolysis represents the optimum conditions for the formation of nanocubes. These nanocubes reflect the basic cubic unit cell of Cu_2O.

4. Conclusion

Copper oxides, especially cuprous oxide, are of interest because of their applications in solar cell technology. The semiconductor cuprous oxide Cu_2O film has been of considerable interest as a component of solar cells due to its band gap energy and high optical absorption coefficient. Since the properties of cuprous oxide not only depend upon the nature of the material but also upon the way they are synthesized, different methods and results obtained on the synthesis of cuprous oxide by various researchers are discussed in this chapter. The properties of the prepared cuprous oxide films related to surface morphology are presented too. In this chapter, the point is made on electrodeposition of cuprous oxide because electrodeposition techniques are particularly well suited for the deposition of metal oxides with thicknesses down to a few nm. The results obtained show that the cuprous oxide can be used as a potential active material for solar cells application.

5. Acknowledgment

This work was supported by Ministry of Science and Technological Development of Republic of Serbia, Project No. OI 172 060.

6. References

Akimoto, K.; Ishizuka, S.; Yanagita, M.; Nawa, Y.; Paul, G. K. & Sakurai, T. (2006). Thin Film Deposition of Cu$_2$O and Application for Solar Cells. *Solar Energy*, Vol. 80, No.6, (June 2006), pp. 715–722, ISSN 0038-092X

Armelao, L.; Barreca, D.; Bertapelle, M.; Bottaro, G.; Sada, C. & Tondello, E. (2003). A Sol–gel Approach to Nanophasic Copper Oxide Thin Films. *Thin Solid Films*, Vol.442, No.1-2, (October 2003), pp. 48–52, ISSN 0040-6090

Briskman, R.N. (1992). A Study of Electrodeposited Cuprous Oxide Photovoltaic Cells. *Solar Energy Materials and Solar Cells*, Vol.27, No.4, (September 1992), pp. 361–368, ISSN 0927-0248

Bugarinović, S.J.; Grekulović, V.J.; Rajčić-Vujasinović, M.M.; Stević, Z.M. & Stanković, Z.D. (2009). Electrochemical Synthesis and Characterization of Copper(I) Oxide (in Serbian). *Hemijska Industrija*, Vol.63, No.3, (May-June 2009), pp. 201-207, ISSN 0367-598X

Daltin, A-L.; Addad, A. & Chopart, J-P. (2005). Potentiostatic Deposition and Characterization of Cuprous Oxide Films and Nanowires. *Journal of Crystal Growth*, Vol.282, No.3-4, (September 2005), pp. 414-420, ISSN 0022-0248

Fortin, E. & Masson, D. (1982). Photovoltaic Effects in Cu$_2$O-Cu Solar Cells Grown by Anodic Oxidation. *Solid-State Electronics*, Vol.25, No.4, (April 1982), pp. 281-283, ISSN 0038-1101

Georgieva, V. & Ristov, M. (2002). Electrodeposited Cuprous Oxide on Indium Tin Oxide for Solar Applications. *Solar Energy Materials and Solar Cells*, Vol.73, No.1, (May 2002), pp. 67–73, ISSN 0927-0248

Ghosh, S.; Avasthi, D.K.; Shah, P.; Ganesan, V.; Gupta, A.; Sarangi, D.; Bhattacharya, R. & Assmann, W. (2000). Deposition of Thin Films of Different Oxides of Copper by RF Reactive Sputtering and Their Characterization. *Vacuum*, Vol.57, No.4, (June 2000), pp. 377-385, ISSN 0042-207X

Golden, T.D.; Shumsky, M.G.; Zhou, Y.; Vander Werf, R.A.; Van Leeuwen, R.A. & Switzer, J.A. (1996). Electrochemical Deposition of Copper (I) Oxide Films. *Chemistry of Materials*, Vol. 8, No.10, (October 1996), pp. 2499–2504, ISSN 0897-4756

Grozdanov, I. (1994). Electroless Chemical Deposition Technique for Cu$_2$O Thin Films. *Materials Letters*, Vol.19, No.5-6, (May 1994), pp. 281–285, ISSN 0167-577X

Han, K. & Tao, M. (2009). Electrochemically Deposited p-n Homojunction Cuprous Oxide Solar Cells. *Solar Energy Materials and Solar Cells*, Vol. 93, No.1, (January 2009), pp.153-157, ISSN 0927-0248

Hu, F.; Chan, K.C. & Yue, T.M. (2009). Morphology and Growth of Electrodeposited Cuprous Oxide under Different Values of Direct Current Density. *Thin Solid Films*, Vol. 518, No.1, (November 2009), pp. 120–125, ISSN 0040-6090

Jayatissa, A.H.; Guo, K. & Jayasuriya, A.C. (2009). Fabrication of Cuprous and Cupric Oxide Thin Films by Heat Treatment. *Applied Surface Science*, Vol.255, No.23, (September 2009), pp. 9474-9479, ISSN 0169-4332

Kobayashi, H.; Nakamura , T. & Takahash, N. (2007). Preparation of Cu_2O Films on MgO (110) Substrate by Means of Halide Chemical Vapor Deposition under Atmospheric Pressure. *Materials Chemistry and Physics*, Vol.106, No.2-3, (December 2007), pp. 292-295, ISSN 0254-0584

Kunze, J.; Maurice, V.; Klein, L.H.; Strehblow, H.H. & Marcus, P. (2003). In Situ STM Study of the Effect of Chlorides on the Initial Stages of Anodic Oxidation of Cu(111) in Alkaline Solutions. *Electrochimica Acta*, Vol.48, No.9, (April 2003), pp.1157-1167, ISSN 0013-4686

Lee, Y.H.; Leu, I.C.; Chang, S.T.; Liao, C.L. & Fung, K.Z. (2004). The Electrochemical Capacities and Cycle Retention of Electrochemically Deposited Cu_2O Thin Film Toward Lithium. *Electrochimica Acta*, Vol.50, No.2-3, (November 2004), pp. 553–559, ISSN 0013-4686

Lincot, D. (2005). Electrodeposition of semiconductors. *Thin Solid Films*, Vol. 487, No.1-2, (September 2005), pp. 40–48, ISSN 0040-6090

Liu, Y.L.; Liu, Y.C.; Mu, R.; Yang, H.; Shao, C.L.; Zhang, J.Y.; Lu, Y.M.; Shen, D.Z. & Fan, X.W. (2005). The Structural and Optical Properties of Cu_2O Films Electrodeposited on Different Substrates. *Semiconductor Science and Technology*, Vol.20, No.1, pp. 44-49, (January 2005), ISSN 0268-1242

Mahalingam, T.; Chitra, J.S.P.; Rajendran, S.; Jayachandran, M. & Chockalingam, M.J. (2000). Galvanostatic deposition and characterization of cuprous oxide thin films. *Journal of Crystal Growth*, Vol. 216, No.1-4, (June 2000), pp. 304–310, ISSN 0022-0248

Markworth, P.R.; Liu, X.; Dai, J.Y.; Fan, W.; Marks, T.J. & Chang, R.P.H. (2001). Coherent Island Formation of Cu_2O Films Grown by Chemical Vapor Deposition on MgO (110). *Journal of Materials Research*, Vol.16, No.8, (August 2001), pp. 2408-2414, ISSN 0884-2914

Maruyama, T. (1998). Copper Oxide Thin Films Prepared from Copper Dipivaloylmethanate and Oxygen by Chemical Vapor Deposition. *Japanese Journal of Applied Physics*, Vol. 37, No.7A, pp. 4099-4102, ISSN 0021-4922

Medina-Valtierra, J.; Ramırez-Ortiz, J.; Arroyo-Rojas, V.M.; Bosch, P. & De los Reyes, J.A. (2002). Chemical Vapor Deposition of $6CuO \cdot Cu_2O$ Films on Fiberglass. *Thin Solid Films*, Vol. 405, No.1-2, (February 2002), pp. 23–28, ISSN 0040-6090

Mizuno, K.; Izaki, M.; Murase, K.; Shinagawa, T.; Chigane, M.; Inaba, M.; Tasaka, A. & Awakura, Y. (2005). Structural and Electrical Characterizations of Electrodeposited p-type Semiconductor Cu_2O Films. *Journal of The Electrochemical Society*, Vol.152, No.4, pp. C179–C182, ISSN 0013-4651

Mukhopadhyay, A.K.; Chakraborty, A.K.; Chatterjee, A.P. & Lahiri, S.K. (1992). Galvanostatic Deposition and Electrical Characterization of Cuprous Oxide Thin Films. *Thin Solid Films*, Vol.209, No.1, (March 1992), pp. 92-96, ISSN 0040-6090

Musa, A.O.; Akomolafe, T. & Carter, M.J. (1998). Production of Cuprous Oxide, a Solar Cell Material, by Thermal Oxidation and a Study of Its Physical and Electrical Properties. *Solar Energy Materials and Solar Cells*, Vol.51, No.3-4, (February 1998), pp. 305-316, ISSN 0927-0248

Nair, M.T.S.; Guerrero, L.; Arenas, O.L. & Nair, P.K. (1999). Chemically Deposited Copper Oxide Thin Films: Structural, Optical and Electrical Characteristics. *Applied Surface Science*, Vol.150, No.1-4, (August 1999), pp. 143-151, ISSN 0169-4332

Nozik, A.J. (1978). Photoelectrochemistry: Applications to Solar Energy Conversion. *Annual Review of Physical Chemistry*, Vol.29, No.1, (October 1978), pp. 189-222, ISSN 0066-426X

Ottosson, M.; Lu, J. & Carlsson, J-O. (1995). Chemical Vapour Deposition of Cu_2O on MgO(100) from CuI and N_2O: Aspects of Epitaxy. *Journal of Crystal Growth*, Vol.151, No.3-4, (June 1995) pp. 305-311, ISSN 0022-0248

Ottosson, M. & Carlsson, J-O. (1996). Chemical Vapour Deposition of Cu_2O and CuO from CuI and O_2 or N_2O. *Surface and Coatings Technology*, Vol. 78, No.1-3, (January 1996), pp. 263-273, ISSN 0257-8972

Papadimitropoulos, G.; Vourdas, N.; Vamvakas, V.Em. & Davazoglou, D. (2005). Deposition and Characterization of Copper Oxide Thin Films. *Journal of Physics: Conference Series, Vol. 10*, No.1, pp. 182–185, ISSN 1742-6588

Pollack, G.P. & Trivich, D. (1975). Photoelectric Properties of Cuprous Oxide. *Journal of Applied Physics*, Vol.46, No.1, (January 1975), pp. 163–172, ISSN 0021-8979

Rai, B.P. (1988). Cu_2O Solar Cells: A Review. *Solar Cells*, Vol.25, No.3, (December 1988), pp. 265–272, ISSN 0379-6787

Rajčić-Vujasinović, M.; Stević, Z. & Djordjević, S. (1994). Application of Pulse Potential for Oxidation of Natural Mineral Covellite (in Russian). *Zhurnal prikladnoi khimii (Russian Journal of Applied Chemistry)*, Vol.67, No.4, pp. 594-597, ISSN 1070-4272

Rajčić-Vujasinović, M.M.; Stanković, Z.D. & Stević, Z.M. (1999). The Consideration of the Analogue Electrical Circuit of the Metal or Semiconductor/Electrolyte Interfaces Based on the Time Transient Analysis. *Elektrokhimiya (Russian Journal of Electrochemistry)*, Vol.35, No.3, pp. 347-354, ISSN 1023-1935

Rakhshani, A.E.; Al-Jassar, A.A. & Varghese, J. (1987). Electrodeposition and Characterization of Cuprous Oxide. *Thin Solid Films*, Vol.148, No.2, (April 1987), pp. 191-201, ISSN 0040-6090

Rakhshani, A.E. & Varghese, J. (1987). Galvanostatic Deposition of Thin Films of Cuprous Oxide. *Solar Energy Materials*, Vol.15, No.4, (May-June 1987), pp. 237-248, ISSN 0165-1633

Ray, S.C. (2001). Preparation of Copper Oxide Thin Film by The Sol–Gel-Like Dip Technique and Study of Their Structural and Optical Properties. *Solar Energy Materials and Solar Cells*, Vol.68, No.3-4, (June 2001), pp. 307-312, ISSN 0927-0248

Rizzo, G.; Arsie, I. & Sorrentino, M. (October 2010). Hybrid Solar Vehicles, In: *Solar Collectors and Panels, Theory and Applications*, Manyala, R. (Ed.), pp. 79-96, Sciyo, ISBN 978-953-307-142-8, Available from:
http://www.intechopen.com/articles/show/title/hybrid-solar-vehicles

Santra, K., Sarkar, C.K.; Mukherjee, M.K & Ghosh, B. (1992). Copper Oxide Thin Films Grown by Plasma Evaporation Method. *Thin Solid Films*, Vol. 213, No.2, (June 1992), pp.226-229, ISSN 0040-6090

Santra, K.; Chatterjee, P. & Sen Gupta, S.P. (1999). Powder Profile Studies in Electrodeposited Cuprous Oxide Films. *Solar Energy Material and Solar Cells*, Vol. 57, No.4, (April 1999), pp. 345–358, ISSN 0927-0248

Sears, W.M. & Fortin, E. (1984). Preparation and properties of Cu_2O/Cu photovoltaic cells. *Solar Energy Materials*, Vol.10, No.1, (April-May 1984), pp. 93-103, ISSN 0165-1633

Singh, D.P.; Singh, J.; Mishra, P.R.; Tiwari, R.S. & Srivastava, O.N. (2008). Synthesis, characterization and application of semiconducting oxide (Cu_2O and ZnO) nanostructures. *Bulletin of Materials Science*, Vol.31, No.3, (June 2008), pp. 319–325, ISSN 0250-4707

Siripala, W., Perera, L.D.R.D., De Silva, K.T.L.; Jayanetti, J.K.D.S. & Dharmadasa, I.M. (1996). Study of Annealing Effects of Cuprous Oxide Grown by Electrodeposition Technique. *Solar Energy Materials and Solar Cells*, Vol.44, No.3, (November 1996), pp. 251–260, ISSN 0927-0248

Stanković, Z.; Rajčić-Vujasinović, M.; Vuković, M.; Krčobić, S. & Wragg, A.A. (1998). Electrochemical Synthesis of Cupric Oxide Powder. Part I: Influence of pH. *Journal of Applied Electrochemistry*, Vol.28, No.12, (December 1998), pp. 1405-1411, ISSN 0021-891X

Stanković, Z.; Rajčić-Vujasinović, M.; Vuković, M.; Krčobić, S. & Wragg, A.A. (1999). Electrochemical Synthesis of Cupric Oxide Powder. Part II: Process Conditions. *Journal of Applied Electrochemistry*, Vol.29, No.1, (January 1999), pp. 81-85, ISSN 0021-891X

Stević, Z. & Rajčić-Vujasinović, M. (2006). Chalcocite as a Potential Material for Supercapacitors. *Journal of Power Sources*, Vol.160, No.2, (October 2006), pp. 1511-1517, ISSN 0378-7753

Stević, Z.; Rajčić-Vujasinović, M. & Dekanski, A. (2009). Estimation of Parameters Obtained by Electrochemical Impedance Spectroscopy on Systems Containing High Capacities. *Sensors*, Vol.9, No.9, (September 2009), pp. 7365-7373, ISSN 1424-8220

Stević, Z. & Rajčić-Vujasinović, M. (In press). Supercapacitors as a Power Source in Electrical Vehicles, in: *Electric Vehicles / Book 1*. Soylu, S. (Ed). ISBN 978-953-307-287-6

Tang, Y.; Chen, Z.; Jia, Z.; Zhang, L. & Li, J. (2005). Electrodeposition and Characterization of Nanocrystalline Films Cuprous Oxide Thin Films on TiO_2. *Materials Letters*, Vol.59, No.4, (February 2005), pp. 434–438, ISSN 0167-577X

Vieira, J.A.B. & Mota, A.M. (2010). Maximum Power Point Tracker Applied in Batteries Charging with Photovoltaic Panels, In: *Solar Collectors and Panels, Theory and Applications*, Ochieng, R.M. (Ed.), pp. 211-224, Sciyo, ISBN 978-953-307-142-8, Available from: http://www.intechopen.com/articles/show/title/maximum-power-point-tracker-applied-to-charging-batteries-with-pv-panels

Wang, L.C.; de Tacconi, N.R.; Chenthamarakshan, C.R.; Rajeshwar, K. & Tao, M. (2007). Electrodeposited Copper Oxide Films: Effect of Bath pH on Grain Orientation and Orientation-dependent Interfacial Behavior. *Thin Solid Films*, Vol. 515, No.5, (January 2007), pp. 3090–3095, ISSN 0040-6090

Wijesundera, R.P., Hidaka, M., Koga, K., Sakai, M. & Siripala, W. (2006). Growth and Characterization of Potentiostatically Electrodeposited Cu_2O and CuO Thin Films. *Thin Solid Films*, Vol.500, No.1-2, (April 2006), pp. 241–246, ISSN 0040-6090

Wong, E.M. & Searson, P.C. (1999). ZnO Quantum Particle Thin Films Fabricated by Electrophoretic Deposition. *Applied Physics Letters*, Vol.74, No.20, (May 1999), pp. 2939-2941, ISSN 0003-6951

Xue, J. & Dieckmann, R. (1990). The Non-Stoichiometry and the Point Defect Structure of Cuprous Oxide ($Cu_{2-\delta}O$). *Journal of Physics and Chemistry of Solids*, Vol. 51, No.11, pp. 1263-1275, ISSN 0022-3697

Zhou, Y. & Switzer, J.A. (1998). Electrochemical Deposition and Microstructure of Copper (I) Oxide Films. *Scripta Materialia*, Vol.38, No.11, (May 1998), pp. 1731-1738, ISSN 1359-6462

10

Semiconductor Superlattice-Based Intermediate-Band Solar Cells

Michal Mruczkiewicz, Jarosław W. Kłos and Maciej Krawczyk

Faculty of Physics, Adam Mickiewicz University, Poznań
Poland

1. Introduction

The efficiency of conversion of the energy of photons into electric power is an important parameter of solar cells. Together with production costs, it will determine the demand for the photovoltaic device and its potential use (Messenger & Ventre, 2004). The design of artificial nanostructures with suitably adjusted properties allows to increase the performance of solar cells. The proposed concepts include, among others, third-generation devices such as tandem cells, hot carrier cells, impurity photovoltaic and intermediate-band cells (Green, 2003). In this chapter we discuss the theoretical model of intermediate-band solar cell (IBSC), the numerical methods of determining the band structure of heterostructures, and the latest reported experimental activities. We calculate the efficiency of IBSCs based on semiconductor superlattices. The detailed balance efficiency is studied versus structural and material parameters. By adjusting these parameters we tailor the band structure to optimize the efficiency.

The background of the concept of IBSC lies in the impurity solar cell concept proposed by (Wolf, 1960) and presented in Fig. 1. The idea was to increase the efficiency by the introduction of intermediate states within a forbidden gap of the semiconductor. This allows the absorption of low-energy photons and causes them to contribute to the generated photocurrent via two-photon absorption. However, as shown experimentally by (Guettler & Queisser, 1970), the introduction of intermediate levels via impurities will create non-radiative recombination centers and cause a degradation of the solar cell efficiency. This effect was studied theoretically by (Würfel, 1993) and (Keevers & Green, 1994), with the conclusion that the introduced impurity levels can increase the efficiency in some cases, but only marginally. However the research in this field is still active and recently the optical transition between CB and IB band in the GaN_xAs_{1-x} alloys was proved experimentally (López et al., 2011; Luque, 2011).

Another, more sophisticated approach to the concept of impurity solar cell was proposed by (Barnham & Duggan, 1990). A further discussion in (Araujo & Martí, 1995), (Luque & Martí, 2001), (Martí et al., 2006) led to the conclusion that the problems related to the impurity states in the solar cell concept might be overcome if the impurities interacted strongly enough to form an impurity band (IB). In such conditions the electron wave functions in the IB are delocalized, causing the radiative recombinations to predominate over the non-radiative ones. The efficiency of the system was described by (Luque & Martí, 1997) on the basis of the extended Shockly-Queisser model (Shockley & Queisser, 1961), the most commonly used and described in detail in the next section. Many extended versions of the model have been developed, such as that proposed by (Navruz & Saritas, 2008) in a study of the effect of the absorption coefficient, or the model of (Lin et al., 2009), considering the carrier mobility and recombinations.

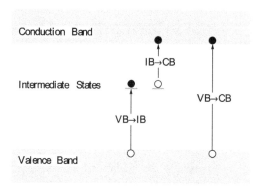

Fig. 1. Model of single-gap solar cell with impurity states introduced. Two possible ways of electron-hole creation are shown: via one-photon absorption in a transition from the valence band to the conduction band (VB→CB), and via two-photon absorption, in which the electron is excited from the valence band to the impurity state (VB→IB) by one photon, and from the impurity state to the conduction band (IB→CB) by another photon.

2. Theoretical model

2.1 Single gap solar cell

Unlike the thermodynamic limits (Landsberg & Tonge, 1980), the limit efficiency considered in the Shockley-Queisser detailed balance model of single-gap solar cell (SGSC) (Shockley & Queisser, 1961) incorporates information on the band structure of the semiconductor and the basic physics. The model includes a number of fundamental assumptions, which allow to evaluate, question and discuss its correctness. All incident photons of energy greater than the energy gap (E_G) of the semiconductor are assumed to participate in the generation of electron-hole pairs. Other assumptions include that no reflection occurs on the surface of the solar cell, the probability of absorption of a photon with energy exceeding the energy gap and creation of electron-hole pair equals one, and so does the probability of collection of the created electron-hole pairs. In the detailed balance model only radiative recombinations between electrons and holes are allowed, by Planck's law proportional to the temperature of the cell. According to this model, all the carriers relax immediately to the band edges in thermal relaxation processes.

The current-voltage equation of the cell under illumination can be written in the following form:

$$J(V) = J_{SC} - J_{Dark}(V),$$ (1)

where J_{SC} is the short circuit current, extracted from the cell when its terminals are closed and the load resistance is zero; the short circuit current is independent of the voltage, but depends on the illumination; the dark current J_{Dark} is the current that flows through the p-n junction under applied voltage, in the case of a solar cell, produced at the terminals of the device under the load resistance R. The detailed balance efficiency is defined as the ratio of the output power P_{out} extracted from the cell to the input power P_{in} of the incident radiation:

$$\eta = \frac{P_{out}}{P_{in}} = \frac{J_m V_m}{P_{in}},$$
(2)

where V_m and J_m is the voltage and current, respectively, that corresponds to the optimal value of the output power.

Both P_{in} and $J(V)$ can be defined in terms of fluxes of absorbed and emitted photons. Let β_s be the incident photon flux, or the number of incident photons per second per square meter received from the sun and the ambient. By Planck's law, describing the blackbody radiation:

$$\beta_s(E) = \frac{2F_s}{h^3 c^2} \frac{E^2}{e^{E/k_b T_a} - 1},$$
(3)

where h is the Planck constant, c is the velocity of light, k_b is a Boltzman constant and T_a is a temperature of the ambient. F_s is a geometrical factor determined by the half of the angle subtended by the sunlight:

$$F_s = \pi \sin^2 \frac{\Theta_{sun}}{2}.$$
(4)

In all the examples discussed in this chapter the maximum concentration of sunlight, corresponding to $\Theta_{sun} = 180°$, is assumed. For that reason there is no need to describe the incident photon flux cming from the ambient and the photon flux described by the equation (3) is the total incident photon flux. The radiation of the sun is coming from all directions. If a flat solar panel receives radiation over a hemisphere, the geometrical factor becomes π, which is equivalent to the cell illuminated with $\Theta_{sun} = 180°$.

The input power will be the total energy of all the incident photons:

$$P_{in} = \int_0^\infty E \beta_s(E) \, dE.$$
(5)

The short circuit current can be expressed as the elementary charge multiplied by the number of absorbed photons, with the absorption coefficient $a(E)$:

$$J_{SC} = q \int a(E) \beta_s(E) \, dE = q \int_{E_G}^\infty \beta_s(E) \, dE,$$
(6)

where the absorption coefficient $a(E)$ (zero for energies lower than the bandgap, one otherwise) determines the lower boundary of the integral.

The dark current is related to the number of photons emitted by the p-n junction:

$$J_{Dark}(V) = q \int e(E) \beta_e(E, V)) \, dE,$$
(7)

where $e(E)$ is an emission coefficient which describe the probability of the photon emission. The generalized form of Planck's law of blackbody radiation (Landau & Lifshitz, 1980) describes the dependence of the flux β_e of photons emitted by the device on the chemical

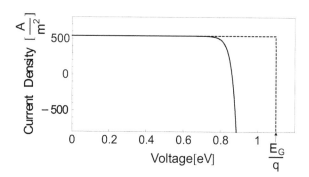

Fig. 2. The current-voltage characteristic of an SGSC with $E_G = 1.1$ eV. The solid and dashed lines represent the $J(V)$ function for a flat cell without concentrators, placed on Earth at a temperature of 300 K and at absolute zero (the temperature corresponding to the ultimate efficiency), respectively.

potential difference, which can be defined by the potential at the terminals:

$$\beta_e(E, \Delta\mu) = \frac{2F_e}{h^3 c^2} \frac{E^2}{e^{(E-\Delta\mu)/k_b T_c} - 1}, \tag{8}$$

where T_c is the temperature of the cell, and $\Delta\mu$ is the chemical potential difference defined as the difference of the quasi-Fermi levels (defined in the next Section):

$$\Delta\mu = E_{FC} - E_{FV} = qV. \tag{9}$$

The lower boundary of the integral (7) depends on the emissivity, $e(E)$ (one for energies above E_G, zero otherwise) of the p-n junction, and thus determines the maximum voltage of the junction (the maximum load resistance that can be applied). Above this voltage the device will emit light.
The current-voltage function (1) becomes:

$$J(V) = q \int_{E_G}^{\infty} (\beta_s(E) - \beta_e(E, V)) \, dE. \tag{10}$$

Figure 2 presents the current-voltage characteristics of a cell with bandgap E_G at different temperatures. As established above, the maximum voltage (at $T = 0$ K) is determined by E_G. In the limit of $T = 0$ K temperature the value of efficiency achieves its maximum value for the specific solar cell, i.e., the ultimate efficiency.

2.2 Intermediate band solar cells
In this section we will show how to extend the expression (10) to the case of the cell with intermediate band. The model IBSC device, shown in Fig. 3, includes emitters n and p, for separation and extraction of the carriers, and an intermediate band (IB) absorber material placed between them. It is desirable that the IB be thermally separated from the valence band (VB) and the conduction band (CB), so that the number of electrons in the IB can only be changed via photon absorption or emission. This assumption allows to introduce three

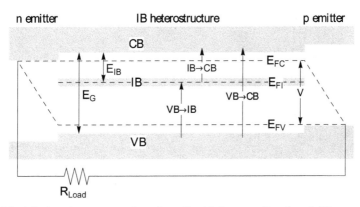

Fig. 3. Model of the band structure of a solar cell with intermediate band. The terminals of the solar cell are connected to the n and p emitters. The possible excitation processes, via one-photon or two-photon absorption, are indicated by arrows. Up down arrows indicate energy differences between band edges.

quasi-Fermi levels, one for each band, to describe the population of electrons within the bands. An infinite mobility of electrons is assumed, to ensure constant quasi-Fermi levels across the junction and minimize the occurrence of non-radiative light traps. The introduction of the IB can improve the efficiency by allowing the absorption of low-energy photons, and thus overcome the problems of the impurity level concept. In Fig. 3 the lowest energy difference between the bands is seen to depend on the value of E_{IB}, the energy difference between the IB and the CB; E_{IB} determines also the threshold energy of the absorbed photons.

In the basic version of the model, the absorption and emission coefficients between each band are assumed to be as presented in Fig. 4. It would probably be more realistic, but still advantageous, to assume that the absorption coefficients corresponding to different transitions are constant, but differ in value. Since the photons that contribute to the transitions between VB and CB predominate in the incident light, the transitions between IB are CB are much weaker that those between VB and IB. According to Martí et al. (2006), the problem has not yet been studied systematically. However, this assumption seems to reflect the behavior of real systems. Thus, the absorption coefficient for different transitions will fulfill the relation:

$$\alpha_{VC} > \alpha_{VI} > \alpha_{IC}. \tag{11}$$

This allows to assume specific values of the absorption coefficients in Fig. 4, but implies that the absorption between IB and CB will be marginal, and so will be the current generated by two-photon absorption.

The assumed form of the absorption and emission functions allows to specify the boundaries of the integrals in the expression for the photon flux absorbed or emitted by the band, analogously to the SGSC model. Three fluxes are distinguished, one for each of the three transitions: VB-CB, VB-IB and IB-CB. Each of the three fluxes contains information on the number of absorbed and emitted photons per unit of time per unit of area:

$$\int_{E_G}^{\infty} (\beta_s(E) - \beta_e(E, \mu)) \, dE, \tag{12}$$

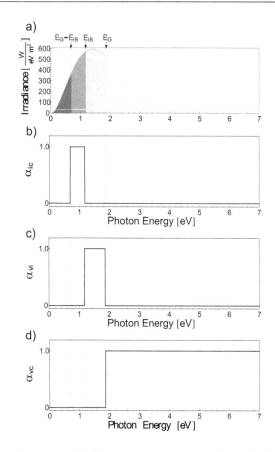

Fig. 4. (a) Radiant emittance of a blackbody at a temperature of 5760 K. Below, plots of the absorption coefficients for (b) IB→CB, (c) VB→IB and (d) VB→CB transitions. The shape of these functions depends on the energy gap and the assumptions made. The depicted forms allow to determine the integral boundaries in equations (12), (13) and (14).

$$\int_{E_G-E_{IB}}^{E_G} (\beta_s(E) - \beta_e(E, \mu_1))\, dE, \tag{13}$$

$$\int_{E_{IB}}^{E_G-E_{IB}} (\beta_s(E) - \beta_e(E, \mu_2))\, dE, \tag{14}$$

where:

$$\mu_1 = E_{FC} - E_{FI}, \tag{15}$$

$$\mu_2 = E_{FI} - E_{FV}. \tag{16}$$

In the equilibrium state the number of electrons in the IB must be constant, which implies that the increase/decrease due to the VB-IB transition must be equal to the decrease/increase due to the IB-CB transition:

$$\int_{E_G-E_{IB}}^{E_G} (\beta_s(E) - \beta_e(E, \mu_1))\, dE = \int_{E_{IB}}^{E_G-E_{IB}} (\beta_s(E) - \beta_e(E, \mu_2))\, dE. \qquad (17)$$

The separation of the quasi-Fermi levels is determined by the applied load resistance and the voltage produced at the terminals of the solar cell:

$$qV = E_{FC} - E_{FV} = (E_{FC} - E_{FI}) + (E_{FI} - E_{FV}) = \mu_1 + \mu_2. \qquad (18)$$

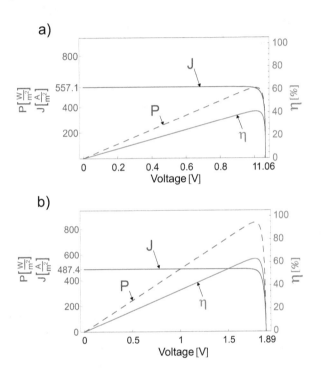

Fig. 5. Voltage dependence of the current density, J, output power P and efficiency η for (a) a single-gap solar cell with $E_G = 1.08$ eV; (b) an intermediate-band solar cell with $E_G = 1.9$ eV, $E_{IB} = 0.69$ eV. The cell has a temperature of 300 K; the incident light is characterized by the blackbody radiation at 5760 K and has a maximum concentration. The band alignment corresponds to the maximum efficiency.

With the last two equations we can calculate the quasi-Fermi level separation for a given voltage (Ekins-Daukees et al., 2005), and thus obtain the current-voltage characteristic. Figure 5 shows the J-V characteristics of (a) an SGSC and (b) an IBSC. The assumed energy gap and intermediate band energy level correspond to the highest possible efficiency of the cell illuminated by sunlight characterized by the 5760 K blackbody radiation, with a maximum concentration. Presented in the same graph, the output power plot shows an increase in efficiency. The short circuit current value is lower in the case of IBSC, but the significant

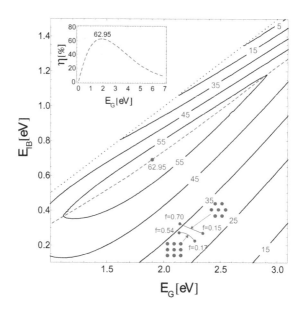

Fig. 6. Contour plot depicting the detailed balance efficiency η versus the energy gap E_G and the distance E_{IB} between the intermediate band bottom and the CB bottom. The values of E_{IB} range from 0 to $\frac{E_G}{2}$. However, according to the model assumed the efficiency is symmetric with respect to E_{IB} in the range from 0 to E_G. the inset in the top-left corner shows the η in dependence on E_G along the dashed line marked in the main figure. The inset shows the changes of E_G, E_I (and η) for AlGaAs supperlattices in dependence on filling fraction (cf. Fig. 11, 12).

increase in the operating voltage leads to a net increase in the efficiency. An explanation of the decrease in the short circuit current in the IBSC (when low-energy photon are absorbed) is provided by Fig. 4, showing the absorption coefficient dependence in the optimal IBSC. The high power absorbed by the cell is seen to contribute to the two-photon processes.

The contour plot in Fig. 6 shows the efficiency versus the bandgap and the distance between IB and CB. These results are important for the understanding of the potential of the IB concept. Later in this chapter they will be compared with simulation data, analyzed in terms of the material parameters used.

If the bandwidth of the solar cell is wider than the distance from the intermediate band to the nearest band, the spectral selectivity might be disturbed. However, these processes are not considered in this chapter. The bandwidth is assumed to only affect the absorption and emission spectra in one of the narrow gaps, changing the boundaries of the integrals in equation (17):

$$\int_{E_G-E_{IB}}^{E_G} (\beta_s(E) - \beta_e(E,\mu_1))\, dE = \int_{E_{IB}-\Delta IB}^{E_G-E_{IB}} (\beta_s(E) - (\beta_e(E,\mu_2)))\, dE, \tag{19}$$

where ΔIB is the intermediate band width.

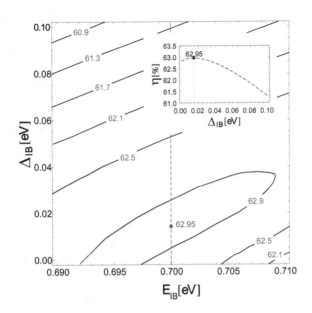

Fig. 7. Detailed balance efficiency of an IBSC with a fixed energy gap $E_G = 1.90$ eV versus the width Δ_{IB} and position E_{IB} of the intermediate band. In the inset, the profile of the efficiency function across the dashed line in the contour plot is shown.

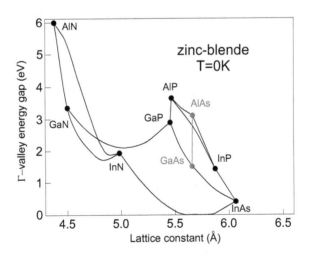

Fig. 8. Energy gap width versus lattice constant for selected III-V group semiconductor compounds. Red line corresponds to the ternary alloy AlGaAs. Note the lattice constant does not change significantly with changing Al concentration in the alloy. (The data have been taken from (Vurgaftman et al., 2001))

In this case an increase in the width of the intermediate band will result in increased absorption, since the gap will shrink. On the other hand, the maximum applicable voltage will decrease with increasing bandwidth, as the emission function will be affected. As shown in Fig. 7, for a fixed energy gap the maximum efficiency may increase. The inset presents the efficiency plotted versus the IB width for energy gap $E_G = 1.90$ eV and intermediate band level $E_{IB} = 0.7$ eV, measured to the top edge of the IB. A maximum is found to occur for $\Delta IB = 0.015$ [eV]; the maximum value is $\eta = 62.96\%$. An improvement by 0.04% is reported in (Green, 2003). An increase by 0.03% is achieved in the cell illuminated by the 6000 K blackbody radiation for $\Delta IB = 0.02$ eV.

3. Calculation of the band structure

We consider a 2D semiconductor superlattice which consists of a periodic array of semiconductor inclusions embedded in a semiconductor matrix. Such a system has an artificially introduced periodicity with a lattice constant much larger than the interatomic distances. As a result of introducing this additional periodicity the conduction and valence bands split into a set of minibands. In this regime of length and energy we can regard the system as continuous on the atomic scale and, in the case of direct gap semiconductors, use the effective parameters describing the position and curvature of the conduction band bottom and the valence band top. Then, the miniband structure of the conduction and valence bands can be calculated with the aid of effective Hamiltonians with spatially dependent effective parameters (Bastart, 1988; Burt, 1999; Califano & Harison, 2000). In the case of semiconductors with a relatively wide gap, such as AlGaAs, the electronic system can be decoupled from the system of light and heavy holes. Also, the stress at the inclusion/matrix interfaces can be neglected in materials of this kind, because of the small atomic lattice constant changes related to the different concentration of Al in the alloy (see Fig. 8). Thus, the simple BenDaniel-Duke Hamiltonian (BenDaniel & Duke, 1966) can be used for electrons in the vicinity of point Γ of the solid semiconductor structure:

$$\left[-\alpha \left(\frac{\partial}{\partial x} \frac{1}{m^*(r)} \frac{\partial}{\partial x} + \frac{\partial}{\partial y} \frac{1}{m^*(r)} \frac{\partial}{\partial y} + \frac{\partial}{\partial z} \frac{1}{m^*(r)} \frac{\partial}{\partial z} \right) + E_C(r) \right] \Psi_e(r) = E\Psi_e(r), \quad (20)$$

where r is the position vector in 3D space. The dimensionless constant $\alpha = 10^{-20} \hbar^2/(2m_e e) \approx 3.80998$ (m_e and e are the free electron mass and charge, respectively) allows to express the energy and the spatial coordinates in eV and Å, respectively; m^* is the effective mass of the electron; E_C denotes the conduction band bottom. Both parameters are periodic with the superlattice period:

$$m^*(\mathbf{r} + \mathbf{R}) = m^*(\mathbf{r}),$$
$$E_C(\mathbf{r} + \mathbf{R}) = E_C(\mathbf{r}), \quad (21)$$

where $\mathbf{R}$ is a lattice vector of the superlattice. We have used the following empirical formulae for a linear extrapolation of the material parameter values in GaAs and AlAs to estimate their values in the $Al_x Ga_{1-x}As$ matrix: $E_C = 0.944x$ and $m^* = 0.067 + 0.083x$, x is a concentration of the Al in GaAs (Shanabrook et al., 1989; Vurgaftman et al., 2001).

In the case of a zinc blende structure (e.g., AlGaAs) both the light- and heavy-hole bands must be taken into account. The Schrödniger equation for each component of the envelope function for light-holes, Ψ_{lh} and heavy-holes Ψ_{hh} reads (Datta, 2005):

$$
-\begin{pmatrix}
\hat{P}+\hat{Q} & 0 & -\hat{S} & \hat{R} \\
0 & \hat{P}+\hat{Q} & \hat{R}^* & \hat{S}^* \\
-\hat{S}^* & \hat{R} & \hat{P}-\hat{Q} & 0 \\
\hat{R}^* & \hat{S} & 0 & \hat{P}-\hat{Q}
\end{pmatrix} \Psi_h(r) = E\Psi_h(r),
\tag{22}
$$

where

$$
\Psi_h(r) = \big(\Psi_{lh\uparrow}(r), \Psi_{lh\downarrow}(r), \Psi_{hh\downarrow}(r), \Psi_{hh\uparrow}(r)\big)^T.
\tag{23}
$$

The subscripts lh and hh label the components of the envelope function for the light and heavy holes, respectively. The symbols $\uparrow$ and $\downarrow$ refer to bands related to opposite z components of the light- and heavy-hole spins. The operators $\hat{P}, \hat{Q}, \hat{R}$ and $\hat{S}$ have the form:

$$
\hat{P} = E_V(r) + \alpha \left(\frac{\partial}{\partial x}\gamma_1(r)\frac{\partial}{\partial x} + \frac{\partial}{\partial y}\gamma_1(r)\frac{\partial}{\partial y} + \frac{\partial}{\partial z}\gamma_1(r)\frac{\partial}{\partial z} \right),
$$

$$
\hat{Q} = \alpha \left(\frac{\partial}{\partial x}\gamma_2(r)\frac{\partial}{\partial x} + \frac{\partial}{\partial y}\gamma_2(r)\frac{\partial}{\partial y} - 2\frac{\partial}{\partial z}\gamma_2(r)\frac{\partial}{\partial z} \right),
$$

$$
\hat{R} = \alpha\sqrt{3}\left[-\left(\frac{\partial}{\partial x}\gamma_2(r)\frac{\partial}{\partial x} - \frac{\partial}{\partial y}\gamma_2(r)\frac{\partial}{\partial y} \right) + i\left(\frac{\partial}{\partial x}\gamma_3(r)\frac{\partial}{\partial y} + \frac{\partial}{\partial y}\gamma_3(r)\frac{\partial}{\partial x} \right) \right],
$$

$$
\hat{S} = \alpha\sqrt{3}\left[\left(\frac{\partial}{\partial x}\gamma_3(r)\frac{\partial}{\partial z} + \frac{\partial}{\partial z}\gamma_3(r)\frac{\partial}{\partial x} \right) - i\left(\frac{\partial}{\partial y}\gamma_3(r)\frac{\partial}{\partial z} + \frac{\partial}{\partial z}\gamma_3(r)\frac{\partial}{\partial y} \right) \right].
\tag{24}
$$

The Luttinger parameters $\gamma_1, \gamma_2, \gamma_3$, describe, the effective masses $1/(\gamma_1 + \gamma_2)$ and $1/(\gamma_1 - \gamma_2)$ of light and heavy holes near point Γ of the atomic lattice are, like the position of the valence band top E_V, periodic in the superlattice structure:

$$
\gamma_\beta(\mathbf{r}+\mathbf{R}) = \gamma_\beta(\mathbf{r}),
$$
$$
E_V(\mathbf{r}+\mathbf{R}) = E_V(\mathbf{r}),
\tag{25}
$$

where the subscript β is 1,2 or 3. For periodic heterostructures consisting of a triangular or square lattice-based system of GaAs rods embedded in $Al_xGa_{1-x}As$, the following material parameter values, dependent on the concentration of Al in aluminium gallium arsenide, can be assumed (Shanabrook et al., 1989; Vurgaftman et al., 2001):

$$
E_V = 1.519 + 0.75x,
$$
$$
\gamma_1 = 6.85 - 3.40x,
$$
$$
\gamma_2 = 2.10 - 1.42x,
$$
$$
\gamma_3 = 2.90 - 1.61x.
\tag{26}
$$

We are interested in the calculation of the spectra of a finite-thickness periodic layer of inclusions (see Fig. 9). In such superlattices, when the superlattice period and the layer thickness are of the order of a few nanometers the lowest miniband within the CB is detached from the other CB minibands. Moreover, the higher CB minibands overlap to form a continuous energy range without minigaps.

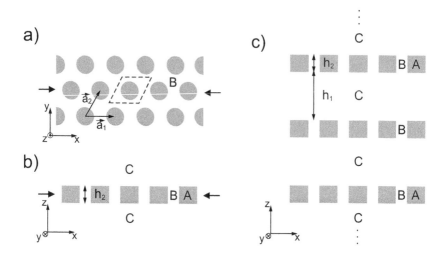

Fig. 9. Structure of a periodic slab with inclusions in triangular lattice, (a) top and (b) side view. Letters A, B and C denote the inclusion, matrix and spacer materials, respectively. The arrows indicate the cross-section plane. Dashed parallelogram in (a) delimits the unit cell, which reproduces the whole plane when translated by superlattice vectors a_1 and a_1. (c) The supercell structure used in the plane wave method: an infinite stack of replicas of the periodic slab.

In the VB all the minibands overlap or are separated by extremely narrow minigaps. Let us assume for simplicity that the total spectrum can by approximated by the model with a single gap (delimited by the top of the highest VB miniband and the bottom of the block of higher CB minibands) and a single intermediate band formed by the first (lowest) CB miniband.

This simplification allows us to calculate the detailed balance efficiency of solar energy conversion for a superlattice-based solar cell using the model with a single intermediate band within the gap.

3.1 Plane wave method

We have calculated the band structure of electrons and holes by the plane wave method (PWM), a technique successfully applied to studying the electronic states in semiconductor heterostructures with quantum dots and wires of different shape and size, as well as interdiffusion and strain effects on electronic bands (Cusack et al., 1996; Gershoni et al., 1988; Li & Zhu, 1998; Li et al., 2005; Ngo et al., 2006; Tkach et al., 2000). By Fourier-expanding the spatially dependent structural parameters: m^*, γ_β, E_C, E_V, and the electron and hole envelope functions the differential equations (20) and (22) can be transformed to a set of algebraic equations for the Fourier coefficients of the envelope functions. This set of equations has the form of an eigenvalue problem with eigenvalues being the energies of successive minibands for the selected wave vector.

The PWM can only be applied to periodic systems. The structure under consideration is finite in one direction, though. To adopt the method to the case considered we calculate the spectrum of an infinite stack of weakly coupled periodic layers, as presented in Fig. 9(c). If

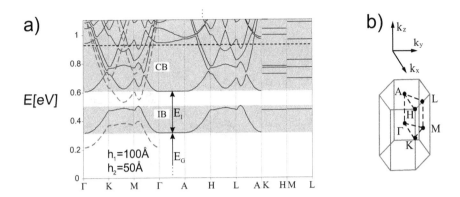

Fig. 10. (a) Electronic minibands in the structure presented in Fig. 9, with GaAs cylinders (material A) embedded in $Al_{0.35}Ga_{0.65}As$ slabs (B) separated by an AlAs spacer (C). Red dashed line represents bands in a 2D superlattice formed by an array of infinitely long rods (i.e., for $k_z = 0$). For a sufficiently thick spacer layer the minigaps are dispersionless in the z direction (lines K-H and M-L in the Brillouin zone shown in (b)). This proves a good separation of the periodic slabs. The calculations were performed for a superlattice with lattice constant $a = 50$ Å and filling fraction $f = 0.3$. The reference energy level $E = 0$ eV corresponds to the CB bottom in solid GaAs. The slab thickness h_2 is 50 Å and the AlAs spacer thickness h_1 is 100 Å.

the distance between adjacent layers is large and the potential in the spacer material C forms a high barrier both for electrons and holes, the spectrum of the system is very close to that of a single isolated layer (Rodríguez-Bolívar et al., 2011).

Figure 10(a) shows the electronic spectrum of the structure presented in Fig. 9, with circular GaAs rods embedded in AlGaAs slabs. Adjacent GaAs/AlGaAs slabs are separated by an AlAs spacer, relatively thick and with a high potential to ensure a good separation of the periodic slabs. This is reflected in the flat dispersion in the z direction (high-symmetry lines K-H, M-L, Γ-A) and the repeated shape of the dispersion branches $\Gamma - K - M - \Gamma$ and $A - H - L - A$. Thus, the case considered proves equivalent to that of a single periodic slab. In the considered range of structural parameter values the electronic spectrum includes one clearly detached miniband and a continuous block of minibands above it. The VB minibands (not shown in Fig. 10) overlap. Thus, the model with a single intermediate band (formed by the first CB miniband) within the energy gap (between the VB and the block of CB minibands) can be used for the calculation of the detailed balance efficiency.

4. Detailed balance efficiency of periodic semiconductor slab

We calculate the electronic and hole spectra of periodic semiconductor layers with different filling fraction values. The filling fraction is defined as the ratio of the in-plane cross-section S_{inc} of the inclusion to the area S of the unit cell area:

$$f = \frac{S_{inc}}{S}. \tag{27}$$

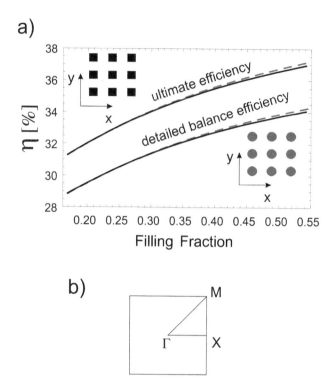

Fig. 11. (a) Detailed balance efficiency and ultimate efficiency of solar energy conversion versus filling fraction for a slab (of thickness h_2=50Å) with cylinders (dashed red line) and square prisms (solid black line) arranged in a square lattice (the lattice constant of the superlattice is $a = 50$ Å). The inclusion, slab and spacer materials are GaAs, $Al_{0.35}Ga_{0.65}As$ and AlAs, respectively. (b) The high-symmetry line in the first Brillouin zone used in the search of absolute minigaps.

We consider two shapes of the inclusions: cylinders and square prisms, and two lattices: the square and triangular lattice. Thus, four combinations of the system geometry are possible. For each combination we calculate the position and width of the valence and conduction bands versus the filling fraction. The following parameters of the band structure are extracted from the calculations:

- the width of the energy gap E_G between the top of the VB and the bottom of the block of CB minibands,

- the shift E_I between the bottom of the first CB miniband and the bottom of the block of higher CB minibands,

- the width ΔE_I of the intermediate band (the first CB miniband).

All three parameters are used in the calculation of the detailed balance efficiency of solar energy conversion for four geometries mentioned above.

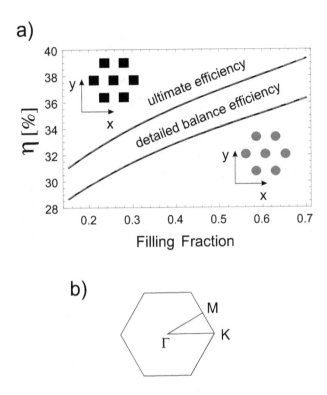

Fig. 12. (a) Detailed balance efficiency and ultimate efficiency of solar energy conversion versus filling fraction for a slab (of thickness h_2=50Å) with cylinders (dashed red line) and square prisms (solid black line) arranged in a triangular lattice (the lattice constant of the superlattice is $a = 50$ Å). The inclusion, slab and spacer materials are GaAs, $Al_{0.35}Ga_{0.65}As$ and AlAs, respectively. (b) The high-symmetry line in the first Brillouin zone used in the search of absolute minigaps.

The lattice constant of the superlattice is fixed at $a = 50$ Å. The assumed thickness of the periodic slab is $h_2 = 50$ Å. A maximum efficiency can be observed in this size range, with the thickness of the periodic slab comparable to the lattice constant of the superlattice. The inclusion and matrix materials are GaAs and $Al_{0.35}Ga_{0.65}As$, respectively. A thick AlAs spacer (of thickness $h_1 = 100$Å) ensures a good separation of adjacent periodic slabs in the PWM supercell calculations. We used 15x15x15 and 13x13x13 plane waves in the calculations of the electronic and hole spectra, respectively.

Figures 11(a) and 12(a) present the calculated ultimate efficiency and detailed balance efficiency versus filling fraction. To investigate the width of the absolute minibands/minigaps we calculated the electronic and hole spectra along the high-symmetry lines shown in Figs.11(b) and 12(b) for square and triangular lattices. The assumed upper bound of ≈ 0.7 of the filling fraction range in Fig. 12 (triangular lattice) corresponds to the maximum filling fraction values, or touching adjacent inclusions, in the considered structures: 0.68 for cylindrical inclusions and 0.86 for inclusions in the shape of square prism. The upper limit of filling fraction for square lattice $\approx$ 0.55 in Fig. 11 results form minibands crossing (and disappearing of the minigap between the lowest conduction miniband and the rest of the minibands in CB) for higher values of f. The lower limit of the filling fraction, $f = 0.15$, is assumed for accuracy of the results obtained by the PWM. A good convergence of results is achieved for intermediate filling fraction values (far from 0 and 1).

The shape of the inclusions is seen to play no important role as long as the filling fraction value is kept. This is evidenced by the overlapping of the solid black and dashed red lines, referring to structures with square prisms and cylinders, respectively, in Figs. 11 and 12. Only in the square lattice case, a minor difference in the efficiency is seen to occur due to the inclusion shape in the range of larger filling fraction values.

Crucial for the efficiency are the lattice symmetry and the filling fraction. In the considered structures the efficiency tends to be higher in the case of the triangular lattice. The maximum values of ultimate (detailed balance) efficiency, 39.5% (35.5%) for the triangular lattice, are reached for large filling fraction values, corresponding to inclusions getting in touch. For this lattice type the efficiency grows monotonically with increasing filling fraction. In the case of the square lattice the filling fraction dependence of the ultimate (detailed balance) efficiency has a maximum of 37% (34%) for filling fraction value $f \approx 0.55$ for which the band crossing appears. For both lattice types the efficiency changes more rapidly in the range of low filling fraction values.

The change of filling fraction affects on the band structure and the values of E_G and E_I parameters which determine the efficiency of solar energy conversion in the supperlatice based solar cell. In the inset in Fig. 6 the detailed balance efficiency as a function of E_G and E_I was plotted (see the curves marked in red) for considered structures. We showed the results for systems with circular inclusions only because the week dependence of η on the inclusion shape. One can notice that the structure (or material) more suitable for solar cell application than the considers supperlattices has to poses slightly smaller energy gap E_G and much bigger separation between IB and CB, E_I. It is evident (see Fig. 8) that materials with better values of the energy gap can be chosen, e.g., InAs/GaAs (Jolley et al., 2010; Zhou et al., 2010) where the lattice mismatch between compounds induces stress which modify the band structure (39% detailed balance efficiency is predicted (Tomić, 2010), other compounds form III-V group (InAs/Al$_x$Ga$_{1-x}$As, InAs$_{1-x}$N$_y$/AlAs$_x$Sb$_{1-x}$, and InAs$_{1-z}$N$_z$/Al$_x$[Ga$_y$In$_{1-y}$]$_{1-x}$P) where 60% detailed balance efficiency at maximum gap concentration where recently calculated (Linares et al., 2011).

5. Experiments

One of the first important experiments related directly to the concept of IBSC was the study by Luque et al. (2005), in which two main operating principles required from the IBSC (the production of photocurrent from photons of energy below the bandgap and the occurrence of three separate quasi-Fermi levels) were confirmed by measurements of quantum efficiency and electroluminescence. In the paper Martí et al. (2006) claim they demonstrate for the first time the production of photocurrent with a simultaneous absorption of two sub-bandgap energy photons, and prove the possibility of obtaining the photocurrent by two-photon absorption processes in InAs/(Al,Ga)As quantum dot (QD) structures.

Although the first experiments prove the possibility of photocurrent production by two-photon absorption, many obstacles still prevent the achievement of satisfactory performance.   In another study, Martí et al. (2007) demonstrate that strain-induced dislocations can propagate from the QD region to the p emitter, resulting in a decrease in the minority carrier lifetime and, consequently, reducing the efficiency. Most of the conclusions drawn from the recent experiments point out the common problem of decrease in open-circuit voltage. These effects are investigated in Jolley et al. (2010).

6. Conclusions

We have examined thoroughly the effect of the (super)lattice symmetry and the cross-sectional shape of the rods on the efficiency of the solar cells based on thin slabs of quantum wire arrays by comparing structures consisting of GaAs cylinders or square prisms, embedded in $Al_xGa_{1-x}As$ and disposed in sites of square or triangular lattice. We show the gain in efficiency of semiconductor slabs with 2D periodicity in comparison with the efficiency of monolithic semiconductors (with a single bandgap). The key role in the gain of efficiency is played by the lowest conduction miniband, which, detached from the overlapping conduction minibands, acts as an intermediate band that opens an extra channel for carrier transitions between the valence band and the conduction band. Another parameter of vital importance for the efficiency of solar-energy conversion is the distance between the top of the highest valence miniband and the bottom of the overlapping conduction minibands. This distance, determines the energy of utilized carriers. Even though the obtained values of the width of the minigap are not optimal to obtain maximal efficiency (Bremner et al., 2009; Luque & Martí, 1997; Shao et al., 2007), the increase in efficiency is significant as compared to the bulk solar cells (Kłos & Krawczyk, 2009).   The impact of the discretisation of the valence band on the efficiency of the photovoltaic effect is be much lesser than that of the discretisation of the conduction band, and, in the first approximation, can be regarded as limited to a shift in valence band top in the evaluation of solar cell efficiency (Kłos & Krawczyk, 2009).

7. References

Araujo, G. L. & Martí, A., (1995). Electroluminescence coupling in multiple quantum well diodes and solar cells, *Appl. Phys. Lett.* Vol. 66: 894Û895.

Ashcroft, N. W. & Mermin, D. N., (1976) *Solid State Physics*, Saunders Collage Publishing.

Barnham, K. W. J. & Duggan, G., (1990). A new approach to high-efficiency multi-band-gap solar cells, *J. Appl. Phys.* Vol. 67: 3490-3493.

Bastard, G. (1988). *Wave Mechanics Applied to Semiconductor Heterostructures*, Les editions de Physique, Paris.

BenDaniel, D. J. & Duke, C. B. (1966). Space-charge effects on electron tunneling, *Phys. Rev.* Vol. 152: 683-692.

Bremner, S. P., Levy, M. Y. & Honsberg, C.B. (2009). Limiting efficiency of an intermediate band solar cell under a terrestrial spectrum, *Appl. Phys. Lett.* Vol. 92: 171110-1-3.

Burt, M.G. (1999). Fundamentals of envelope function theory for electronic states and photonic modes in nanostructures, *J. Phys.: Condens. Matter* Vol. 11, R53-R83.

Califano, M. & Harison, P. (2000). Presentation and experimental validation of a single-band, constant-potential model for self-assembled InAs/GaAs quantum dots, *Phys. Rev. B* Vol. 61: 10959-10965.

Cusack, M. A., Briddon, P. R. & Jaros, M. (1996). Electronic structure of InAs/GaAs self-assembled quantum dots, *Phys. Rev. B* Vol. 54: R2300-R2303.

Datta, S. (2005). *Quantum Transport: Atom to Transistor*, Cambridge University Press, Cambridge.

Ekins-Daukes, N. J., Honsberg, C. B. & Yamaguchi M., (2005). Signature of intermediate band materials from luminescence measurements, *Presented at the 31st IEEE Photovoltaic Specialists Conference*.

Gershoni, D., Temkin, H., Dolan, G. J., Dunsmuir, J., Chu, S. N. G. & Panish, M. B. (1988). Effects of two-dimensional confinement on the optical properties of InGaAs/InP quantum wire structures, *Appl. Phys. Lett.* Vol. 53: 995-997.

Green, M. A. (2003). *Third Generation Photovoltaics: Advanced Solar Energy Conversion*, Springer-Verlag, Berlin.

Guettler, G. & Queisser, H. J. (1970). Impurity photovoltaic effect in silicon, *Energy. Conversion* Vol. 10(2): 51-5.

Jolley, G., Lu, H. F., Fu, L., Tan, H. H. & Jagadish, C. (2010). Electron-hole recombination properties of $In_{0.5}Ga_{0.5}As/GaAs$ quantum dot solar cells and the influence on the open circuit voltage, *Appl. Phys. Lett.* Vol. 97: 123505-1-3.

Keevers, M.J. & Green, M.A., (1994). Efficiency improvements of silicon solar cells by the impurity photovoltaic effect, *J. Appl. Phys.* Vol. 75: 4022-4031.

Kłos, J. W. & Krawczyk, M., (2009). Two-dimensional GaAs/AlGaAs superlattice structures for solar cell applications: Ultimate efficiency estimation, *J. Appl. Phys.* Vol. 106: 093703-1-9.

Kłos, J. W. & Krawczyk, M., (2010). Electronic and hole spectra of layered systems of cylindrical rod arrays: aolar cell application, *J. Appl. Phys.* Vol. 107: 043706-1-5.

Krawczyk, M. & Kłos, J. W., (2010). Electronic and hole minibands in quantum wire arrays of different crystallographic structure, *Physics Letters A* Vol. 374: 647-654.

Landau, L.D. & Lifshitz, E.M., (1980). *Statistical Physics*, Part 1. Vol. 5 (3rd ed.). Butterworth-Heinemann.

Landsberg, P. T. & Tonge, G. (1980). Thermodynamic energy conversion efficiencies, *J. Appl. Phys.* 51: R1-R20.

Li, S.-S. & Zhu, B.-F. (1998). Electronic structures of GaAs/AlAs lateral superlattices, *J. Phys.: Condend. Matter* Vol. 10: 6311-6319.

Li, S.-S., Chang, K. & Xia, J.-B. (2005). Effective-mass theory for hierarchical self-assembly of $GaAs/Al_xGa_{1-x}As$ quantum dots, *Phys. Rev. B* Vol. 71: 155301-1-7.

Lin, A. S., Wang, W. & Phillips, J. D., (2009). Model for intermediate band solar cells incorporating carrier transporta and recombination, *J. Appl. Phys.* Vol. 105: 064512-1-8.

Linares, P. G., Martí, A., Antolín, E. & Luque, A. (2011). III-V compound semiconductor screening for implementing quantum dot intermediate band solar cells, *J. Appl. Phys.* Vol. 109: 014313-1-8.

López, N., Reichertz, L. A., Yu, K. M., Campman, K. & Walukiewicz, W. (2011). Engineering the electronic band structure for multiband solar cells, *Phys. Rev. Lett.* Vol. 106: 028701-1-4.

Luque, A. & Martí, A. (1997). Increasing the efficiency of ideal solar cells by photon induced transitions at intermediate levels, *Phys. Rev. Lett* Vol. 78: 5014-5018.

Luque, A. & Martí, A., (2001). A metallic intermediate band high efficiency solar cell, *Prog. Photovolt. Res. Appl.* Vol. 9(2): 73-86.

Luque, A., Martí, A., López, N., Antolín, E., Cánovas, E., Stanley, C., Farmer, C., Caballero, L. J., Cuadra, L. & Balenzategui, J. L., (2005). Experimental analysis of the quasi-Fermi level split in quantum dot intermediate-band solar cells, *Appl. Phys. Lett.* Vol. 87: 083505-1-3.

Luque, A. & Martí, A., (2011). Towards the intermediate band, *Nature Photonics* Vol. 5: 137-138.

Martí, A., López, N., Antolín, E., Cánovas, E., Stanley, C., Farmer, C., Cuadra, L. & Luque, A., (2006). Novel semiconductor solar cell structures: The quantum dot intermediate band solar cell, *Thin Solid Films* Vol. 511-512: 638-644.

Martí, A., Antolín, E., Stanley, C., Farmer, C., López, N., Díaz, P., Cánovas, E., Linares, P. G. & Luque, A., (2006). Production of photocurrent due to intermediate-to-conduction-band transitions: a demonstration of a key operating principle of the intermediate-band solar cell, *Phys. Rev. Lett.* Vol. 97: 247701-1-4.

Martí, A., López, N., Antolín, E., Cánovas, E., Luque, A., Stanley, C., Farmer, C. & Díaz, P., (2007). Emitter degradation in quantum dot intermediate band solar cells, *Appl. Phys. Lett.* Vol. 90: 233510-1-3.

Messenger, R. A. & Ventre, J. (2004). *Photovoltaic Systems Engineering*, CRC Press, Florida.

Navruz, T. S. & Saritas, M., (2008). Efficiency variation of the intermediate band solar cell due to the overlap between absorption coefficients, *Solar Energy Materials & Solar Cells* Vol. 92: 273-282

Nelson, J. (2003). *The Physics of Solar Cells*, Imperial College, UK.

Ngo, C. Y., Yoon, S. F., Fan, W. J. & Chua, S. J. (2006). Effects of size and shape on electronic states of quantum dots, *Phys. Rev. B* Vol. 74: 245331-1-10.

Rodríguez-Bolívar, S., Gómez-Campos, F. M., Luque-Rodríguez, A., López-Villanueva, J. A., Jiménez-Tejada, J. A. & Carceller, J. E. (2011). Miniband structure and photon absorption in regimented quantum dot systems, *J. Appl. Phys.* Vol. 109: 074303-1-7.

Shanabrook, B. V., Glembocki, O. J., Broido, D. A. & Wang, W. I. (1989). Luttinger parameters for GaAs determined from the intersubband transitions in $GaAs/Al_xGa_{1-x}As$ multiple quantum wells, *Phys. Rev. B* Vol. 39: 3411-3414.

Shao, Q., Balandin, A. A., Fedoseyev, A. I. & Turowski, M. (2007). Intermediate-band solar cells based on quantum dot supracrystals, *Appl. Phys. Lett.* Vol. 91: 163503-1-3.

Shockley, W. & Queisser, H. J. (1961). Detailed balance limit of efficiency of p-n junction solar cells, *J. Appl. Phys.* Vol. 32, 510-519.

Tkach, N. V., Makhanets, A. M. & Zegrya, G. G. (2000). Energy spectrum of electron in quasiplane superlattice of cylindrical quantum dots, *Semicond. Sci. Technol.* Vol. 15: 395-398.

Tomić, S. (2010). Intermediate-band solar cells: Influence of band formation on dynamical processes in InAs/GaAs quantum dot arrays, *Phys. Rev. B* Vol. 82: 195321-1-15.

Vurgaftman, I., Meyer, J. R. & Ram-Mohan, L. R. (2001). Band parameters for III-V compound semiconductors and their alloys, *J. Appl. Phys.* Vol. 89: 5815-5875.

Wolf, M. (1960). Limitations and possibilities for improvements of photovoltaic solar energy converters, *Proc. IRE* Vol. 48: 1246-1263.

Würfel, P. (1993). Limiting efficiency for solar cells with defects from a three-level model, *Sol. Energy Mater. Sol. Cells* Vol. 29: 403-413.

Zhou, D., Sharma, G., Thomassen, S. F., Reenaas, T. W. & Fimland, B. O. (2010). Optimization towards high density quantum dots for intermediate band solar cells grown by molecular beam epitaxy, *Appl. Phys. Lett.* Vol. 96: 061913-1-3.

Permissions

The contributors of this book come from diverse backgrounds, making this book a truly international effort. This book will bring forth new frontiers with its revolutionizing research information and detailed analysis of the nascent developments around the world.

We would like to thank Professor, Doctor of Sciences, Leonid A. Kosyachenko, for lending his expertise to make the book truly unique. He has played a crucial role in the development of this book. Without his invaluable contribution this book wouldn't have been possible. He has made vital efforts to compile up to date information on the varied aspects of this subject to make this book a valuable addition to the collection of many professionals and students.

This book was conceptualized with the vision of imparting up-to-date information and advanced data in this field. To ensure the same, a matchless editorial board was set up. Every individual on the board went through rigorous rounds of assessment to prove their worth. After which they invested a large part of their time researching and compiling the most relevant data for our readers. Conferences and sessions were held from time to time between the editorial board and the contributing authors to present the data in the most comprehensible form. The editorial team has worked tirelessly to provide valuable and valid information to help people across the globe.

Every chapter published in this book has been scrutinized by our experts. Their significance has been extensively debated. The topics covered herein carry significant findings which will fuel the growth of the discipline. They may even be implemented as practical applications or may be referred to as a beginning point for another development. Chapters in this book were first published by InTech; hereby published with permission under the Creative Commons Attribution License or equivalent.

The editorial board has been involved in producing this book since its inception. They have spent rigorous hours researching and exploring the diverse topics which have resulted in the successful publishing of this book. They have passed on their knowledge of decades through this book. To expedite this challenging task, the publisher supported the team at every step. A small team of assistant editors was also appointed to further simplify the editing procedure and attain best results for the readers.

Our editorial team has been hand-picked from every corner of the world. Their multi-ethnicity adds dynamic inputs to the discussions which result in innovative outcomes. These outcomes are then further discussed with the researchers and contributors who give their valuable feedback and opinion regarding the same. The feedback is then collaborated with the researches and they are edited in a comprehensive manner to aid the understanding of the subject.

Apart from the editorial board, the designing team has also invested a significant amount of their time in understanding the subject and creating the most relevant covers. They scrutinized every image to scout for the most suitable representation of the subject and create an appropriate cover for the book.

The publishing team has been involved in this book since its early stages. They were actively engaged in every process, be it collecting the data, connecting with the contributors or procuring relevant information. The team has been an ardent support to the editorial, designing and production team. Their endless efforts to recruit the best for this project, has resulted in the accomplishment of this book. They are a veteran in the field of academics and their pool of knowledge is as vast as their experience in printing. Their expertise and guidance has proved useful at every step. Their uncompromising quality standards have made this book an exceptional effort. Their encouragement from time to time has been an inspiration for everyone.

The publisher and the editorial board hope that this book will prove to be a valuable piece of knowledge for researchers, students, practitioners and scholars across the globe.

List of Contributors

M. Benhaliliba and C.E. Benouis
Physics Department, Sciences Faculty, Oran University of Sciences and Technology Mohamed Boudiaf- USTOMB, POBOX 1505 Mnaouer- Oran, Algeria

K. Boubaker, M. Amlouk and A. Amlouk
Unité de Physique des dispositifs à Semi-conducteurs UPDS, Faculté des Sciences de Tunis, Campus Universitaire 2092 Tunis, Tunisia

Chunfu Zhang, Hailong You and Yue Hao
School of Microelectronics, Xidian University, China

Zhenhua Lin and Chunxiang Zhu
ECE, National University of Singapore, Singapore

Mukesh Kumar Singh
Uttar Pradesh Textile Technology Institute, Souterganj, Kanpur, India

Yunfei Zhou, Michael Eck and Michael Krüger
University of Freiburg/Freiburg Materials Research Centre, Germany

Malina Milanova
Central Laboratory of Applied Physics, BAS, Bulgaria

Petko Vitanov
Central Laboratory of New Energy & New Energy Sources, BAS, Bulgaria

Almantas Pivrikas
Physical Chemistry, Linz Institute for Organic Solar Cells, Johannes Kepler University Linz, Austria School of Chemistry and Molecular Biosciences, Centre for Organic Photonics and Electronics, The University of Queensland, Brisbane, Austria

Seishi Abe
Research Institute for Electromagnetic Materials, Japan

Doo Hyun Park and Bo Young Jeon
Department of Biological Engineering, Seokyeong University, Seoul, Korea

Il Lae Jung
Department of Radiation Biology, Environmental Radiation Research Group, Korea Atomic Energy Research Institute, Daejeon, Korea

Sanja Bugarinović
IHIS, Science and Technology Park "Zemun", Belgrade, Serbia

Mirjana Rajčić-Vujasinović, Zoran Stević and Vesna Grekulović
University of Belgrade, Technical faculty in Bor, Bor, Serbia

Michal Mruczkiewicz, Jarosław W. Kłos and Maciej Krawczyk
Faculty of Physics, Adam Mickiewicz University, Pozna'n, Poland

Printed in the USA
CPSIA information can be obtained
at www.ICGtesting.com
JSHW011425221024
72173JS00004B/682